CW01206747

FRANK THOMPSON'S CATALOGUE OF MODEL CARS

Contents

1 **INTRODUCTION** — *page 4*

2 **LESNEY**

The golden years of Lesney — *page 5*
Models of yesteryear — *page 71*
Kingsize specials — *page 82*
Major packs — *page 93*
Gift sets — *page 95*

3 **DINKY**

Selected specials — *page 99*
Dublo-dinky toys — *page 103*
Dinky specials of the fifties — *page 104*
Dinky specials - the British Army — *page 107*
Winged wonders — *page 109*

4 **CORGI** — *page 111*

Corgi classics — *page 116*

5 **SPOT ON** — *page 117*

6 **LONE STAR** — *page 122*

7 **BRITAINS** — *page 123*

8 **MINIC** — *page 125*

9 **RIO** — *page 127*

KENNETH MASON · 13-14 HOMEWELL · HAVANT · HAMPSHIRE

Frank Thompson's catalogue of MODEL CARS

Introduction

WHO WOULD HAVE THOUGHT that the toy cars our children used to play with only a few years ago would now have become collectors' pieces and, in some cases, be worth a small fortune. But this indeed is what has happened in the world — my world — of die cast toys, and in particular the model cars, buses, trucks — little works of art in themselves — that used to, and still do, delight the heart of the schoolboy collector.

The magic products of Lesney, Dinky, Corgi, Spot On — the mini matchbox marvels that made the spending of the weekly pocket money an exciting venture — are now important names in the field of collecting. So, as an expert and one who recognised the potential value of these toys, it is a delight for me to think that out of the children's world of playthings has grown a sophisticated new hobby, one that is growing daily and one that has in these last few years acquired international status.

For these reasons, and because I have had hundreds of enquiries from people who wanted to buy, sell or have a model valued, I thought the time had come to provide a guide that would give advice, based on a unique knowledge of these die cast models. As one who has researched the subject for more than ten years, and travelled throughout Britain and the continent in search of collections, individual models or even parts of models, I know better than most just how to assess the quality, rarity and market price of practically every make of model car in the world.

I take great pride in having in my possession what is probably the largest individual collection of toys in existence. And because this fact has become known, not only to collectors in this country, but to connoisseurs in all parts of the world, some of whom have visited me in my shop in York, my advice is being increasingly sought. Collectors from America, Holland, France and the Middle East have come to me for valuations. Demand for information has grown to such an extent that this has underlined the need for a directory that would set out in detail the value of each and every model. In other words a collectors' guide that would become a standard reference work. In my research and study of toys from ancient times to the present day, I have found that none is finer or more worthy of collection than those made in the years 1951 to 1971, which I call our Golden Age. This was a period in which I believe toys were at their very best, when production lines were turning out perfect creations in every detail and correct to scale. The characteristics of these early four wheeled vehicles are a joy to behold.

One point that gives emphasis to the growing value of the model car is that some of the toy companies themselves, who have a special division for collectors, have approached me for help because they have been short of many of the first models. So if you began collecting in the fifties and sixties, and have a set or individual models tucked away in the attic you are lucky indeed.

I have travelled hundreds of miles in search of these early models, literally from Lands End to John O'Groats, and at Wick, in the very North of Scotland, I found valuable additions for my collection. Not content with touring Britain, however, I visited Holland and found some fascinating models in the toy shops of Rotterdam and Amsterdam.

So finding the early gems is a question of diligent searching. In particular watch the shops that are having a clear-out sale. It is surprising what they may turn out from their store rooms. Several times I have found a treasure for my collection from old toys a shopkeeper was virtually throwing away. If for example you have in your possession one of the first 1s 6d Lesney toys, you may have a real treasure. There are others of the same vintage. So if you are lucky enough to discover one, examine it carefully until you discover which model it is. The markings, the colour design, the edition number are the points you have to study. Once you have identified it you can value it by consulting this directory. Whether you wish to sell it or make it a valued part of your collection is up to you.

It has been a long but fascinating task this study of the domain of the model car. Some people were a little sceptical of my faith in the future of the model car as a collector's piece but all my forecasts have been proved right. The same dealers who once doubted my wisdom now come to me for advice.

This book has been a labour of love. The compilation of these many hundreds of prices and descriptions of models has been a loving but painstaking task. Without the help of my wife Anne I doubt if I could have achieved it for she was a constant source of inspiration and help.

NOTE Where a price is not shown against a particular model this indicates that, at time of going to press, the model was available at current list price.

The golden years of Lesney

The firm of Lesney was founded by John W Odell and Leslie G Smith in 1947. They produced small diecast cars as a sideline to their electrical products business. As demand for model cars and trucks outstripped that for the electrical goods, they had to convert the business to toy making only. One of their outstanding early models was the state processional coach with a team of eight horses. The miniature Coronation coach, made in 1953, was another outstanding model. A series called 'The models of yesteryear' brought Lesney to the height of success and made the name a household one in the diecast models world.

1, Road roller. 2, Dumper  3, Cement mixer. 4, M Harris tractor

CURRENT MARKET PRICES AND THE VARIATIONS OF ALL MODELS IN CORRECT COLLECTORS' ORDER

	Mint boxed	Mint unbxd	Good cond	Fair cond
No 1 First issue road roller Green, red metal rollers, tan driver, long flat canopy gold trim, tow hook no number cast on body. Made 1953	£5	£3	£2	£1
1a Light green, metal rollers, crimpled axle	£4	£3	£2	£1
1b Medium green, metal rollers, crimpled axle	£3	£2	£1	50p
1c Dark green, metal rollers, crimpled axle	£5	£4	£3	£2
No 1 Second issue road roller Green, red metal rollers, tan driver, tall peaked canopy, gold trim, tow hook, no number cast on body. Made 1955	£5	£3	£2	£1
1a Light green, light tan driver, metal rollers, crimpled axle	£4	£3	£2	£1
1b Light green, dark tan driver, metal rollers, crimpled axle	£3	£2	£1	50p
No 1 Third issue road roller Green, red metal rollers, green driver, tall peaked canopy, no trim, tow bar, number cast on inside body. Made 1958	£3	£2	£1	50p

	Mint boxed	Mint unbxd	Good cond	Fair cond
1a Light green, metal rollers, crimpled axle		£2	£1	50p
1b Dark green, metal rollers, crimpled axle		£2	£1	50p
1c Dark green, metal rollers, round axle		£2	£1	50p
No 1 Fourth issue road roller Green, red plastic rollers, green driver, tall peaked canopy, no trim, tow bar, number case inside body. Made 1958	£3	£2	£1	50p
1a 'Made in England by Lesney' cast on base	£2	£1	50p	25p
1b 'Made in England' cast under canopy, two rivets and hole in base	£2	£1	50p	25p
No 1 Fifth issue Mercedes truck Light green, removeable canopy, silver grill and half base, tow hook black plastic wheels. Made 1968	£1.50	£1	50p	25p
1a Black plastic wheels, orange canopy	£1.50	£1	50p	25p
1b Licence plate under front bumper	£1.50	£1	50p	25p

6 THE GOLDEN YEARS OF LESNEY

	Mint boxed	Mint unbxd	Good cond	Fair cond
1c Licence plate, tow slot	£1	75p	50p	25p
1d Gold body, orange canopy, large superfast wheels	£1.50	£1	50p	25p
1e Gold body, yellow canopy, large superfast wheels	£1.50	£1	75p	25p
1f Gold body, yellow canopy, small superfast wheels	£1	75p	50p	25p
No 1 Sixth issue Mod Rod Yellow and silver, tinted windshield, silver motor, metal base, sticker decal on hood. Made 1972	50p	40p	25p	25p
Flaw: 3 red wheels, 1 black		£10	£5	
1a Small front and large rear wheels, all superfast, black	50p	40p	25p	25p
1b Small front and large rear wheels, all superfast, red	50p	40p	25p	25p
No 2 Dumper Second model Green body, red dumper, thin tan driver, gold trim, metal wheels, no number cast on body. Made 1953	£5	£3	£2	£1
2a Green painted wheels, crimpled axle, nail holding dumper to body	£4	£3	£2	£1
2b Unpainted metal wheels, crimpled axle, nail holding dumper to body	£4	£3	£2	£1
2c Unpainted metal wheels, crimpled axle, crimpled nail holding dumper to body	£3	£2	£1	50p
No 2 Second issue Dumper Green, red dumper, thick driver, no trim with number cast under bumper at front. Made 1957	£4	£3	£2	£1
2a Metal wheels, crimpled axle	£3	£2	£1	50p
2b Green painted wheels, crimpled axle	£3	£2	£1	50p
2c Green painted wheels, round axle	£3	£2	£1	50p
2d Green painted wheels, no hat on driver	£5	£4	£3	£2
No 2 Third issue, Muirhill Dumper Green, red, 'Laing' decal on doors, black plastic wheels. Made 1961-62	£3	£2	£1	50p
No 2 Fourth issue, Mercedes trailer Light green, tow hook with rotating tow bar, black plastic wheels. Made 1968	£1.50	£1	50p	25p
2a Orange canopy	£1.50	£1	50p	25p

	Mint boxed	Mint unbxd	Good cond	Fair cond
2b Orange canopy, brace inside rear corners	£1	75p	50p	25p
2c Gold body, orange canopy, large superfast wheels	£1.50	£1	50p	25p
2d Yellow canopy, large superfast wheels	£1	75p	50p	25p
No 2 Fifth issue Jeep Hot Rod Pink body, off white seats, silver motor, black exhaust pipes, light green base, superfast wheels. Made 1972	£1	75p	50p	25p
2a Model with large superfast wheels added. Made 1973	£1	75p	50p	25p
No 3 Cement Mixer Third model Blue, orange metal wheels, four paddles inside barrel, no number cast on body. Made 1953	£5	£4	£2	£1
3a Handle on rear door, metal wheels, crimpled axle	£5	£4	£2	£1
3b No handle; metal wheels, crimpled axle	£5	£4	£2	£1
3c Large grey plastic wheels, crimpled axle, no paddles inside body	£4	£3	£2	£1
3d Large grey plastic wheels, round axle with no paddles inside barrel	£3	£2	£1	50p
No 3 Second issue Bedford Tipper Truck Red cab and chassis, brown tipper, number cast under cab. Made 1961	£4	£3	£2	£1
3a Grey cab, maroon tipper	£4	£3	£2	£1
3b Grey cab, grey plastic wheels, knobby treads, no trim	£3	£2	£1	50p
3c Black plastic wheels, knobby treads, silver trim	£3	£2	£1	50p
3d Black plastic wheels, fine treads, no trim	£3	£2	£1	50p
3e Red tipper, grey plastic wheels, fine treads, no trim	£3	£2	£1	50p
3f Red tipper, black plastic wheels, fine treads, no trim	£3	£2	£1	50p
No 3 Third issue Mercedes Ambulance Off white, red cross on door, blue tinted windows, blue dome, white plastic stretcher, black plastic wheels. Made 1968	£5	£4	£2	£1
Flaw: No decal on right hand side	£15			

THE GOLDEN YEARS OF LESNEY

	Mint boxed	Mint unbxd	Good cond	Fair cond
3a Light cream body, sticker decals	£4	£3	£1	50p
3b Rich cream body, sticker decals	£4	£3	£1	50p
3c Off white body, superfast wheels, sticker decals	£4	£3	£1	50p
3d Cream body, superfast wheels, sticker decals	£4	£3	£1	50p
No 4 Massey Harris Tractor Fourth model Red, tan driver, cast to steering wheel, no number, four spokes on front wheels, gold trim and hubs. Made 1954	£5	£3	£2	£1
4a 'Lesney' cast to inside right fender, 'England' cast to inside left fender, metal wheels, crimpled axle	£5	£3	£2	£1
4b 'England' cast to inside right fender, 'Lesney' cast to inside left fender, metal wheels, crimpled axle	£5	£3	£2	£1
4c 'Lesney' cast to inside both fenders, metal wheels, crimpled axle	£5	£3	£2	£1
4d 'England' cast to inside both fenders, metal wheels, crimpled axle	£5	£3	£2	£1
No 4 Second issue Massey Harris Tractor Red, tan driver attached to rear axle, no fenders, number cast inside body, solid front wheels and hollow rear wheels, no trim. Made 1957	£4	£3	£2	£1
4a Bronze hubs, metal wheels, crimpled axle	£4	£3	£2	£1
4b Gold hubs, metal wheels, crimpled axle	£4	£3	£2	£1
4c Grey plastic wheels, crimpled axle	£3	£2	50p	25p
4d Grey plastic wheels, round axle	£3	£2	50p	25p
No 4 Third issue Triumph motor cycle and sidecar Silver blue, silver wire wheels, BP tyres. Made 1959-60	£10	£5	£4	£1
4a Fender connected part way with knobby treads	£10	£5	£4	£1
4b Fender connected in full, fine treads	£10	£5	£4	£1
4c Blocks connecting side car to cycle on base	£10	£5	£4	£1
No 4 Fourth issue State Truck Yellow cab and chassis, green tinted windows with metal base, grill and headlights, black plastic wheels. Made 1967	£1.50	£1	50p	25p
Flaw: 5 four-spoked wheels, 1 eight-spoked wheel	£25			
4a, b, c, d dark grey, light blue, blue green, green stake body	£1.50	£1	50p	25p
4e green stake body with tow slot	£1.50	£1	50p	25p
4f superfast wheels with four spokes on hubs	£1.50	£1	50p	25p
4g superfast wheels with eight spokes on hubs	£1.50	£1	50p	25p
No 4 Fifth issue Gruesome Twosome Superfast wheels, gold with silver trim. Made 1972-75	35p	25p	15p	10p
No 5 First issue London Bus Red with gold radiator. No 5 on front and rear. No number on base. Made 1954. Very rare in mint cond	£10	£5	£3	£2
1a 'Buy Matchbox series' sticker on sides, small recessed square, metal wheels, recess at lower right gear, crimpled axle	£10	£5	£3	£2
1b 'Buy Matchbox series' sticker on sides, no recessed square, metal wheels, crimpled axle	£10	£5	£3	£2
No 5 Second issue London Bus Red with gold radiator. Made 1957. 5 cast on front & rear, no number on base. Very rare	£10	£7	£5	£3
2a 'Buy Matchbox series' decal on side in bright yellow with green letters, metal wheels, crimpled axle	£10	£8	£5	£3
2b 'Buy Matchbox series' decal in dull yellow and green letters on side, metal wheels, crimpled axle	£10	£8	£5	£3
2c 'Buy Matchbox series' on side, grey plastic wheels, silver radiator, crimpled axle	£10	£5	£3	£2
2d 'Buy Matchbox series' decal on side, grey plastic wheels, silver radiator, round axle. Getting harder to find	£10	£5	£3	£2
2e 'Players please' decal on side, grey plastic wheels, round axle	£10	£5	£3	£2
2f 'Players please' decal on side, grey plastic wheels, silver radiator, round axle	£10	£5	£3	£2
No 5 Third issue London Bus Red, cast on base instead of body, baseplate reads in three lines 1961, silver radiator	£8	£4	£2	£1

8 THE GOLDEN YEARS OF LESNEY

5, London bus. 6, Quarry truck
7, Horse drawn milk float. 8, Caterpillar tractor

	Mint boxed	Mint unbxd	Good cond	Fair cond
3a 'Players please' decal on side, grey plastic wheels, round axle	£8	£4	£2	£1
3b 'Visco static' decal, grey plastic wheels, round axle	£8	£4	£2	£1
3c 'Drink Peardrax' decal, grey plastic wheels, round axle	£7	£3	£2	£1
3d 'Drink Peardrax' decal, black plastic wheels, round axle	£7	£4	£3	£2
3e 'Visco static' decals, black plastic wheels, round axle	£7	£4	£3	£2
No 5 Fourth issue London Bus Red & white, plastic seats, baseplate reading in five lines 1965-66, 'Visco static' decals on sides, black plastic wheels	£10	£7	£5	£3
4a 'Longlife' decals, black plastic wheels	£8	£5	£3	£2
4b 'Visco static' decals, 4 line base, black plastic wheels	£8	£5	£3	£2
4c 'Visco static' decals, 4 line base, shallow cut-out under axle, black plastic wheels	£10	£8	£5	£3
4d 'Visco static' decals, black plastic wheels, 4 line base with full cut-out under axle	£8	£5	£3	£1
4e Black plastic wheels, 'Visco static' decals, four line base, with very small support in front of the front axle,	£10	£7	£5	£3
No 5 Fifth issue Lotus Europa Off-white interior, clear windows, 55-1 in blue, superfast wheels. 73mm. Made 1970.	£2	£1	75p	50p
5a Superfast wheels, blue sticker decals	£2	£1	75p	50p

	Mint boxed	Mint unbxd	Good cond	Fair cond
5b Pink, superfast wheels	£2	£1	75p	50p
5c Pink, superfast wheels, sticker decals	£2	£1	75p	50p
5d Pink, superfast, wide racing wheels, decals	£3	£2	£1	75p
No 6 First issue Quarry Truck Orange, grey dumper, six vertical ribs, gold trim, six metal wheels. TT scale 108:1, Made 1954	£6	£5	£4	£3
1a Orange, grey dumper, metal wheels, crimpled axle	£6	£5	£4	£3
No 6 Second issue Euclid Quarry Truck Yellow, four vertical ribs, 6 black plastic wheels, decals on doors. Made 1957	£5	£4	£3	£2
2a Crimpled axle, black plastic wheels with knobby rims	£5	£4	£3	£2
2b Black plastic wheels with knobby rims and round axle	£5	£4	£3	£2
2c Black plastic wheels with fine rims and round axle	£5	£4	£3	£2
No 6 Third issue 10 Wheel Quarry Truck Yellow, silver radiator, black plastic wheels, rear dual wheels, letter A or B found cast under base. 70mm. 80/100 scale. Made 1964	£5	£4	£2	£1
3b Flat front wheels, otherwise as above	£5	£4	£2	£1
3a Recessed rear wheels, rounded front wheels	£5	£4	£2	£1
3c No recess in rear wheels, no silver trim	£5	£4	£2	£1
3d Small step on right front fender, otherwise as 3c	£5	£4	£2	£1
No 6 Fourth issue Ford Pickup Red body, white canopy, cream interior, black plastic wheels, 70 mm 67-1. Made 1969	£3	£2	£1	75p
4a Red body, white canopy, white plastic grill	£3	£2	£1	75p
4b White plastic grill, letter A in front of rear axle, otherwise as above	£3	£2	£1	75p
4c red body, white canopy, silver grill, black plastic wheels	£3	£2	£1	75p
4d White plastic grill, superfast wheels	£2	£1	75p	50p

THE GOLDEN YEARS OF LESNEY 9

	Mint boxed	Mint unbxd	Good cond	Fair cond
4e White plastic grill, green base, superfast wheels	£2	£1	75p	50p
4f Silver plastic grill, green base, superfase wheels	£2	£1	75p	50p
4g Silver plastic grill, wide racing superfast wheels	£2	£1	75p	50p
Flaw: No canopy	£5			

No 6 Fifth issue Mercedes Tourer Orange, black hood, white trim, superfast wheels, 76 mm 58-1. Made 1974

	35p	30p	25p	15p

No 7 First issue Horse drawn milk float Orange, white driver, white bottles, brown horse with white mane and hoofs, gold collar, spoked metal wheels, no number. Rare model. Made 1954

	Mint boxed	Mint unbxd	Good cond	Fair cond
	£25	£20	£10	£7
1a Driver has silver jacket and silver bottles. Metal wheels, crimpled axle. One of the rarest in the world	£1000	£990	£250	£100
1b White bottles, dark brown horse, metal wheels, crimpled axle	£35	£30	£20	£10
1c Light brown horse, metal wheels, crimpled axle	£30	£25	£15	£10
1d Orange cart, orange driver with white hat, grey plastic wheels, crimpled axle	£25	£20	£15	£10
1e Grey plastic wheels, round axles, brown hoofs on horse	£25	£20	£15	£10
1f Orange driver, orange hat, brown hoofs on horse, grey plastic wheels, round axles. Very scarce model	£50	£45	£25	£15

No 7 Second issue Ford Anglia Light blue, clear plastic windows, silver radiator, headlights, bumpers, red tail lights, grey plastic wheels. Made 1961

	£5	£4	£2	£1
2a Light blue, grey plastic wheels with knobby treads, 7 on raised platform of baseplate. Colours can vary	£5	£4	£2	£1
2b Blue or grey, blue bumper, tail lights, windows tinted green, 7 on raised platform, grey plastic wheels with fine treads	£5	£4	£2	£1
2c Silver plastic wheels with knobby treads	£5	£4	£2	£1
2d Rear bumper unpainted, knobby treads and silver plastic wheels	£5	£4	£2	£1
2e Fawn or blue; silver plastic wheels with fine treads	£5	£4	£2	£1
2f Silver plastic wheels, fine treads, platform not raised	£6	£5	£3	£2
2g Fawn, blue or grey, black plastic wheels with fine treads, 7 not on raised platform	£6	£5	£3	£2
2h Fawn, green, blue or brown, black plastic wheels with fine treads, 7 found on raised platform	£5	£4	£2	£1

No 7 Third issue Ford refuse truck Red and grey, 76 mm, 85-1, grey plastic bumper, silver grey metal loader, green tinted windows, black plastic wheels. Made 1967

	£7	£6	£4	£2
3a Orange and grey, straight arm above rear wheels	£7	£5	£3	£2
3b Orange and grey, cutout above rear wheels, black plastic wheels	£7	£5	£3	£1
3c Red or orange, cutout above rear wheels, small tow slot	£7	£5	£3	£2
3d Cutout above rear wheels, superfast wheels and tow slot	£6	£5	£3	£2
3e Yellow/orange body, cutout above rear wheels, tow slot, superfast wheels	£5	£4	£2	£1
3f Yellow/orange, very wide superfast racing wheels	£5	£4	£2	£1

No 7 Fourth issue Hairy Hustler Dark red or pink plastic silver trim, yellow plastic window, yellow interior, 73mm, 52-1, superfast wheels, no 5 Decals on sides and front

	35p	30p	25p	15p

No 8 First issue caterpillar tractor Yellow with rubber treads and no number, tow hook and driver. 'TT' scale 102-I. Made 1955

	£10	£7	£5	£3
1a Orange with gold trim, black hat on driver, metal rollers, crimpled axle, no number, tow hook	£10	£7	£5	£3
1b Yellow, silver trim, metal rollers, crimpled axle	£10	£7	£5	£3

No 8 Second issue caterpillar tractor Golden yellow with 8 on bottom, large stack made 1959

	£10	£8	£6	£4
2a As above with crimpled axle, metal rollers	£10	£8	£6	£4

THE GOLDEN YEARS OF LESNEY

	Mint boxed	Mint unbxd	Good cond	Fair cond
No 8 Third issue caterpillar tractor Yellow, angled tow hook. Length 4.8 cms, TT scale. Made 1961-2	£10	£8	£5	£3
3a Yellow with 8 on bottom, metal rollers, round axle	£8	£6	£4	£2
3b Yellow with 18 on bottom, metal rollers, round axles	£20	£18	£10	£5
3c Yellow, orange with 18 on bottom, metal rollers, round axle	£20	£15	£10	£5
3d Yellow, large silver plastic rollers, 8 on base	£10	£8	£6	£4
3e Yellow, large black plastic rollers, 8 on base	£10	£8	£6	£4
No 8 Fourth issue caterpillar tractor Dark yellow, no driver, green rubber treads, Length 70 mm, 102-1 TT scale. Made 1964	£5	£4	£3	£2
4a Black plastic rollers, 18 misprint on bottom	£15	£5	£3	£2
4b Dark yellow, 8 on bottom, black plastic rollers	£5	£4	£3	£2
4c Dark yellow, thick stack, 8 on bottom	£5	£4	£3	£2
4d Dark yellow, small thin stack, 8 on bottom	£6	£3	£2	£1
No 8 Fifth issue Ford Mustang fastback White, red interior, clear windows, silver hubs, black plastic tyres, length 73mm, 65-1. Made 1967	£4	£3	£2	£1
5a White, silver grill, patent pending on base	£4	£3	£2	£1
5b White, white grill, with patent number, tow slot	£4	£3	£2	£1
5c Orange and red, red interior	£5	£4	£3	£2
5d White, red interior, superfast wheels	£3	£2	£1	50p
5e Orange and red, red interior, superfast wheels	£3	£2	£1	50p
5f Orange and red, white interior, superfast wheels	£3	£2	50p	25p
5g Red, white interior, superfast wheels	£3	£2	50p	25p
No 8 Sixth issue Wildcat dragster Orange, yellow interior, green tinted windows, Black Cat decals on doors, length 73 mm, 65-1. Made 1971	£1	75p	50p	25p
6a Black base, **6b** yellow base, **6c** metal base	£1	75p	50p	25p
Flaw: No label on left side	£10			
No 8 Seventh issue De Tomaso Pantera White, pink interior, 8 decals in black on bonnet and doors, orange recess in boot, length 76 mm, 56-1 Made 1975.				
No 9 First issue Dennis fire escape engine Red, no number, gold trim on radiator, driver's hat, booster under ladder, ladder railings with top dash, HO TT scale 93:1. Made 1956	£25	£9	£5	£3
1a Gold trim, 10 mm wheel at rear of ladder, metal wheels, crimpled axle	£25	£9	£5	£3
1b Gold trim, 13 mm wheel at rear of ladder, metal wheels, crimpled axle	£25	£9	£5	£3
1c No trim on railings, 13 mm wheel at rear of ladder, metal wheels, crimpled axle	£25	£9	£5	£3
1d Gold trim, 9 on front bumper, 13mm wheel at rear of ladder; metal wheels, crimpled axle	£25	£9	£5	£3

9, Fire engine. 10, Mechanical horse and trailer

THE GOLDEN YEARS OF LESNEY 11

	Mint boxed	Mint unbxd	Good cond	Fair cond

1e No trim on railings, 13 mm wheel at rear of ladder, metal wheels, crimpled axle — £25 £9 £5 £3

1f 13 mm wheel at rear of ladder, grey plastic wheels, crimpled axle. Very rare model — £100 £50 £15 £10

1g 13 mm wheel at rear of ladder, grey plastic wheels, round axle. Very rare model — £100 £50 £15 £10

No 9 Second issue Merryweather fire engine Red, 9 on base, yellow ladder on top, HO TT scale. Made 1959/60 — £10 £5 £3 £1

2a Tan ladder, 17 rungs, grey plastic wheels, crimpled axle, silver rear hose connections — £5 £4 £2 £1

2b Tan ladder, 17 rungs, grey plastic wheels, round axle, silver rear hose connections — £5 £4 £2 £1

2c Gold ladder, 17 rungs, grey plastic wheels, round axle, red rear hose connections — £5 £4 £3 £1

2d Gold ladder, 17 rungs, black plastic wheels, knobby treads — £4 £3 £2 £1

2e Silver ladder, 17 rungs, black plastic wheels, knobby treads — £4 £3 £2 £1

2f Silver ladder, 17 rungs, black plastic wheels, fine treads — £4 £3 £2 £1

2g Gold ladder, 18 rungs, black plastic wheels, fine treads — £4 £3 £2 £1

No 9 Third issue boat & trailer Blue trailer, black plastic wheels, white hull on boat with blue deck, clear plastic windscreen, length 89 mm, 125/1. Made 1967 — £4 £3 £2 £1

3a No side supports on boat **3b** Side supports on boat — £4 £2 50p 25p

3c Side supports on boat, superfast wheels — £4 £2 50p 25p

3d Side supports on boat, superfast wheels, very dark blue trailer — £5 £2 £1 50p

No 9 Fourth issue Javelin Green, silver trim on bonnet, orange interior, opening doors, superfast wheels, length 79 mm 63-1. Made 1972 — £2 £1 50p 25p

4a Green, black grill on bonnet — £1 75p 50p 25p

No 9 Fifth issue AMX Javelin Gold, gold interior, black grill on bonnet, superfast wheels, length 76 mm, 63-1. Made 1974 — 50p 25p

5a Dark green, orange interior, black top on bonnet, towing hook for trailers, opening doors, superfast wheels, length 77 mm, 66-1. Made 1976

No 10 First issue mechanical horse and trailer Red cab, three metal wheels, grey trailer with two metal wheels, no number, gold trim, 150:1. Made 1955 — £25 £10 £4 £2

1a Very dark grey trailer, metal wheels, crimpled axle — £10 £5 £2 £1

1b Light grey trailer, metal wheels, crimpled axle — £10 £5 £2 £1

No 10 Second issue mechanical horse and trailer Red cab, tan trailer, no number, gold trim. Made 1958/59 — £25 £10 £5 £2

2a Metal wheels, crimpled axle, gold trim — £10 £5 £3 £2

2b Dark tan trailer, silver trim, metal wheels in front, trailer with grey plastic wheels and crimpled axle — £10 £5 £2 £1

2c Metal wheels in front, grey plastic wheels in rear, round axle, dark tan trailer with grey plastic wheels, round axle, silver trim — £10 £5 £3 £1

No 10 Third issue sugar container truck Blue, Tate and Lyle decals in yellow, red lettering for part of decal, 8 plastic wheels, length 67 mm, TT scale 112-1. Made 1961 — £25 £15 £5 £3

3a Blue, grey plastic wheels or black plastic wheels, decals upside down. A genuine rare fault — £250 £100 £20 £5

3b Grey plastic wheels, knobby treads, very thin support on base between 2nd and 3rd axle — £25 £10 £5 £3

3c Silver plastic wheels, knobby treads, thin support on base between 2nd and 3rd axle — £25 £10 £5 £3

3d Silver plastic wheels, fine treads, thick support on base between 2nd and 3rd axle — £25 £10 £5 £3

THE GOLDEN YEARS OF LESNEY

	Mint boxed	Mint G unbxd	Good cond	Fair cond
3e Grey plastic wheels, fine treads with hole behind second axle in base plate	£25	£10	£5	£3
3f Very dark body, blue headlights, grey plastic wheels, fine treads	£25	£15	£10	£5
3g Black plastic wheels, knobby treads, silver headlights	£25	£15	£10	£5
3h Black plastic wheels, fine treads, silver headlights	£25	£15	£10	£5
3i Black plastic wheels, fine treads, blue headlights	£25	£15	£10	£5
No 10 Fourth issue pipe truck Red, grey plastic pipes; blue tinted windows, black plastic wheels, length 75 mm, 92-1. Made 1967	£5	£4	£2	£1
4a 'Leyland' silver grill, and 'Leyland' under base	£4	£3	£1	50p
4b Ergonmatic cab on base, silver grill	£4	£3	£1	50p
4c White plastic grill, base with tow slot	£4	£3	£1	50p
4d Silver base, grill with tow slot	£4	£3	£1	50p
4e Silver base, grill, superfast wheels, tow slot	£4	£3	£1	50p
4f Light orange body, yellow pipes, silver base, grill, superfast wheels, tow slot	£4	£3	£1	50p
4g Light orange body, yellow pipes, silver base, grill, superfast wheels, tow slot	£3	£2	£1	50p
Flaw: Copper colour one side, orange on other	£25			
No 10 Fifth issue piston popper Blue or purple, silver engine mounted through bonnet, orange interior, silver trim, pacemaker that can be propelled, length 73 mm, 65-1	75p	50p	40p	20p
5a Royal blue, yellow interior, 75 mm	50p	30p	20p	10p
No 11 First issue road tanker Radiator with silver or gold trim, headlights, gas tanks and tank covers on top of body, no number on base. Made 1955.	£25	£10	£5	£3
1a Green with gold trim, metal wheels, crimpled axle, small flat base between cab and body	£25	£10	£5	£3
1b Butterscotch, metal wheels, crimpled axle, silver trim	£25	£10	£5	£3

	Mint boxed	Mint unbxd	Good cond	Fair cond
1c Light yellow, silver or gold trim, metal wheels, crimpled axle	£20	£10	£8	£5
1d Light yellow body, metal wheels, crimpled axle, half round brace between cab and body	£20	£10	£8	£5
1e Red, gold trim, tiny Esso decal at rear, crimpled axle	£30	£10	£7	£5
1f Red, gold or silver trim, metal wheels, large Esso decal at rear, crimpled axle	£25	£15	£10	£5
No 11 Second issue road tanker All bodies were red in this issue; silver trim on radiator, tank covers on top of body with number on base, TT scale 100:1, Made 1958/59	£15	£10	£7	£5
2a Half round base, metal wheels, crimpled axle	£10	£7	£5	£3
2b Red gas tanks, half round base, metal wheels, crimpled axle	£10	£7	£5	£3
2c Red gas tanks, half round base, grey plastic wheels, crimpled axle	£10	£7	£5	£3
2d Red gas tanks, half round base, grey plastic wheels, round axle	£10	£7	£5	£3
2e Small flat support on base, hole in base, grey plastic wheels, round axle	£10	£7	£5	£3
2f Half round base, black plastic wheels, round axle, knobby treads	£10	£7	£5	£3
No 11 Third issue Jumbo crane Yellow, red hydraulic lifter sleeve, black plastic wheels, length 76 mm, 84-1, HO scale. Made 1965	£5	£4	£2	£1
3a All yellow, double cable hook, open brace at top of boom, small front wheels	£5	£4	£3	£2
3b All yellow, slightly larger front wheels	£5	£4	£3	£2
3c Red weight box and top brace filled in on boom	£4	£3	£2	£1
3d Red weight box, large single cable hook	£4	£3	£2	£1
3e Red weight box, yellow single cable hook. All the above models in this range come with hooks in reverse positions also	£4	£3	£2	£1

THE GOLDEN YEARS OF LESNEY 13

	Mint boxed	Mint unbxd	Good cond	Fair cond
No 11 Fourth issue scaffolding truck Silver body, red base and grill, green tinted windows, builders supply company decals on sides, yellow or orange plastic scaffolding, length 67 mm, 91-1, scaffolding can be taken from truck and erected like the real thing. Made 1970	£5	£4	£2	£1
4a Truck as above but with decals missing on sides	£15	£10	£5	£3
4b Black plastic wheels, **4c** has superfast wheels, otherwise details as above	£4	£3	£2	£1
No 11 Fifth issue Flying Bug All red body, red fins, yellow jets on roof, driver with silver helmet, black and white decals on bonnet, length 73 mm, 59-1. Made 1973	50p	40p	30p	10p
5a Lighter colour, slightly lighter body weight. Made 1976				
No 12 First issue Land Rover Dull olive green, tan driver, silver trim on headlights and bumper, no number on model. Made 1955	£10	£5	£3	£2
1a Small brace platform between the front axles, metal wheels, crimpled axle	£10	£5	£3	£2
1b Much lighter green body, platform filled in, metal wheels, crimpled axle	£10	£7	£5	£3
No 12 Second issue Land Rover Olive green body, no driver, tow hook with number on silver trim, length 51 mm. Made 1958-9	£10	£5	£3	£1
2a Black plastic wheels, knobby treads, crimpled axle	£7	£4	£2	£1
2b Black plastic wheels, knobby treads, round axle	£7	£4	£2	£1
2c Large grey plastic wheels, fine treads, round axle	£5	£3	£2	£1
2d Black plastic wheels, fine treads, round axle	£5	£3	£2	£1
2e Black plastic wheels, fine treads, round axle and no silver trim	£5	£4	£3	£2

The horn buttons on above models come in varied sizes

	Mint boxed	Mint unbxd	Good cond	Fair cond
No 12 Third issue Safari Land Rover Green in various shades, white plastic interior, tan luggage on roof, clear plastic windows, black plastic wheels, length 70 mm, 72-1.	£6	£5	£3	£2
3a Brown luggage normal	£5	£4	£3	£2
3b Green, brown luggage, small support behind tow hook	£6	£5	£2	£1
3c Light blue body, brown luggage	£5	£4	£2	£1
3d Medium blue body, small support	£4	£3	£2	£1
3e Medium blue, no small support	£4	£3	£2	£1
3f Very dark blue body, tan luggage, no small support, tow hook	£5	£3	£2	£1
3g Darker blue body, tan luggage, small support, tow hook	£5	£3	£2	£1
3h Metallic gold, tan luggage, small support, superfast wheels, tow hook	£4	£3	£2	£1

11, Road tanker. 12, Land Rover

14   THE GOLDEN YEARS OF LESNEY

	Mint boxed	Mint unbxd	Good cond	Fair cond

No 12 Fourth issue Setra Coach Metallic gold lower half, white plastic top, clear windows, metal base and grill, rear plastic lights, superfast wheels, length 76 mm, 100-1. Made 1971 — £5 £4 £2 £1

4a Shades vary, tan upper roof, thin rear reinforcement on base. — £5 £4 £2 £1

4b Shades vary, lower half in various metallic shades, tan upper roof, thick rear reinforcement on base. — £5 £4 £2 £1

4c White upper roof, thin rear reinforcement on base — £5 £4 £3 £2

4d Various shades, white upper roof, thick rear reinforcement on base — £5 £4 £2 £1

4e Yellow lower body, white upper roof. Rare model — £10 £7 £5 £3

No 12 Fifth issue Big Bull Yellow, silver trim on body, black tracks, green shovel, red wheels. Made 1975

No 13 First issue wreck truck Tan, red metal crane and hook, crane attached to body just above rear axle, sides of truck have short curve, silver trim, TT scale 100:1. Made 1955 — £15 £10 £6 £4

1a Metal wheels, crimpled axle, brace between cab and body extends part way — £15 £10 £6 £4

1b Metal wheels, crimpled axle, brace extends to top. Very rare model — £100 £50 £15 £10

No 13 Second issue wreck truck Tan or orange body, red metal crane and hook, crane attached to body axle, sides of truck curve slightly longer than that of first issue, silver trim with number and brace between cab and body extending to top. Made 1958 — £25 £15 £10 £5

2a Metal wheels, crimpled axle — £15 £10 £7 £5

2b Grey plastic wheels, crimpled axle — £15 £10 £7 £5

2c Grey plastic wheels, round axle — £10 £7 £5 £3

No 13 Third issue Thames wreck truck Red body, yellow decal with black, red and green lettering, silver trim with number, length 54mm, TT scale 100-1, metal wheels. matchbox garages. Made 1961 — £15 £10 £5 £3

3a Open lattice on crane, red painted hook, red letters on decal outlined in black, grey plastic wheels, round axle — £10 £5 £3 £2

3b Letters outlined in black, silver metal hook, grey plastic wheels, open lattice, round axle — £10 £5 £3 £2

3c Letters outlined in black, small metal hook, black plastic wheels with open lattice, knobby treads — £10 £5 £3 £2

3d Letters outlined in black, silver metal hook, black plastic wheels with closed lattice on crane, fine treads on wheels — £7 £5 £3 £1

3e Letters not outlined, grey plastic hook, black plastic wheels, fine treads, closed lattice — £5 £4 £2 £1

3f Letters not outlined, grey plastic hook, one parking lamp under right headlight, black plastic wheels, fine treads, closed lattice — £70 £50 £3 £2

13, Wreck truck. 14, Ambulance

THE GOLDEN YEARS OF LESNEY 15

	Mint boxed	Mint unbxd	Good cond	Fair cond
3g Letters not outlined, grey plastic hook, parking lamps under both headlights, black plastic wheels, fine treads, closed lattice	£5	£4	£3	£2

No 13 Fourth issue Dodge wrecker truck Yellow cab with green body, plastic hook BP decals on sides, black plastic wheels, round axle, length 76mm, 85:1, HO scale. Made 1965

	Mint boxed	Mint unbxd	Good cond	Fair cond
	£7	£5	£3	£2
4a Green cab, yellow body, grey plastic hook	£5	£4	£2	£1
4b Yellow cab, green body, grey plastic hook, black plastic wheels	£5	£4	£2	£1
4c Red plastic single cable hook	£5	£4	£2	£1
4d Red plastic double cable hook	£5	£4	£2	£1
4e Red plastic single cable hook with sticker decals	£4	£3	£2	£1
4f Sticker decals, plastic hooks on above models can be changed by hand, superfast wheels	£4	£3	£2	£1

No 13 Fifth issue Baga Buggy Metallic green body, black and red trim, flower decal on bonnet, silver engine in back, superfast wheels, length 67 mm, 49-1. Made 1972

	Mint boxed	Mint unbxd	Good cond	Fair cond
	75p	50p	25p	15p

No 13 Sixth issue Baja Dune Buggy Bright green, silver engine, red trim, red flowered decal on bonnet, superfast wheels, length 66 mm, 49-1. Made 1975

No 14 First issue ambulance Cream body, silver trim, no number, ambulance case on sides, metal wheels, crimpled axle, TT scale 100:1. Made 1956

	Mint boxed	Mint unbxd	Good cond	Fair cond
	£15	£10	£7	£5
1a Metal wheels, crimpled axle	£15	£10	£7	£5
1b Metal wheels, crimpled axle, red cross painted on roof	£15	£10	£7	£5

No 14 Second issue Daimler ambulance Cream and silver, no number, ambulance cast on sides, red cross on roof. Made 1958

	Mint boxed	Mint unbxd	Good cond	Fair cond
	£15	£10	£7	£5
2a Metal wheels, crimpled axle	£10	£7	£5	£3

	Mint boxed	Mint unbxd	Good cond	Fair cond
2b Grey plastic wheels, crimpled axle. Very scarce model	£100	£50	£15	£10
2c Grey plastic wheels, round axle. Rare model	£100	£50	£15	£10
2d Off white or dull yellow, grey plastic wheels, round axle	£15	£10	£7	£5

No 14 Third issue Bedford ambulance Off white, silver trim, red cross on sides, LCC ambulance decals in black letters, interior, length 67 mm, 81-1. Made 1962

	Mint boxed	Mint unbxd	Good cond	Fair cond
	£15	£7	£5	£3
3a White interior, grey plastic wheels, crimpled axle	£10	£8	£7	£3
3b White interior, grey plastic wheels, round axle	£7	£5	£3	£2
3c Cream interior, red cross outline on roof, silver plastic wheels, round axle	£5	£4	£3	£2
3d White interior, red cross outline on roof, silver plastic wheels, round axle	£5	£4	£3	£2
3e Cream interior, no outline on roof, black plastic wheels with knobby treads	£5	£4	£3	£2
3f White interior, no outline on roof, black plastic wheels with knobby treads	£7	£5	£3	£2
3g Blue lettered decal, black plastic wheels, fine treads	£7	£5	£3	£2
3h Dark blue, black lettered decal, black plastic wheels, fine treads	£10	£7	£5	£3

No 14 Fourth issue Iso Grifo Blue with tow hook, clear windows, silver grill, silver headlights and bumper, metal base, length 76 mm, 61-1. Made 1968

	Mint boxed	Mint unbxd	Good cond	Fair cond
	£4	£3	£2	£1
4a Dark blue interior, silver plastic hubs, black plastic wheels	£4	£3	£2	£1
4b Various shades of colour, superfast wheels	£3	£2	£1	50p
4c Light blue interior, superfast wheels	£3	£2	£1	50p
4d White interior, light blue body, superfast wheels	£3	£2	£1	50p
4e Light blue body, wide racing wheels, white interior, superfast wheels	£3	£2	£1	50p

No 14 Fifth issue Mini-ha-ha Dark red, silver trim, blue interior, driver with red helmet,

THE GOLDEN YEARS OF LESNEY

	Mint boxed	Mint unbxd	Good cond	Fair cond

RAF type decals on doors, silver mounted engine on bonnet, length 60mm, 53:1. Made 1976

No 15 First issue truck tractor Silver trim, silver tank behind cab, tow hook and no number. Made 1956 — £10 £5 £3 £2

1a Yellow, six metal wheels, crimpled axle — £10 £5 £3 £2

1b Orange or brown, six metal wheels, crimpled axle — £10 £5 £3 £2

1c Orange, ten grey plastic wheels, crimpled axle — £10 £5 £3 £2

No 15 Second issue Atlantic tractor super Orange with silver trim, black plastic wheel behind cab and tow hook, also with yellow trim, 138:1. Made 1960 — £10 £7 £5 £3

2a Black plastic wheels, round axle, knobby treads, silver or yellow trim — £10 £7 £5 £3

No 15 Third issue refuse truck Blue body, grey dumper, six black plastic wheels, decal on side, length 64 mm, 96-1, TT scale. Made 1964 — £7 £5 £3 £2

3a Silver grill and headlights, knobby treads, decal cut on angle at rear — £7 £5 £3 £2

3b Fine treads otherwise details similar, but note on this model there are many variations of decals, braces and trim — £5 £4 £3 £2

3c No trim and thick hinge inside dump body, decal cut square with brace between gas tanks, fine treads — £7 £5 £3 £2

No 15 Fourth issue Volkswagen 1500 White with black numbered decal 137 on doors, black plastic wheels, silver plastic hubs, length 72 mm, 59-1. Made 1969 — £5 £4 £2 £1

4a Black plastic wheels with decals, details as above — £5 £4 £2 £1

4b White and off white body, sticker decals — £5 £4 £2 £1

4c White and off white body, superfast wheels with decals, superfast wheels, sticker decals — £4 £3 £2 £1

4d Red metallic body, superfast wheels. Many variations and shades and size of numbers — £4 £3 £2 £1

No 15 Fifth issue fork lift truck Red body, yellow and grey lifting device, decals on side in black, red and yellow, length 70 mm, 62-1. Made 1973 — 50p 40p 30p 20p

5a Length 70 mm, with special action feature. Made 1976

No 16 First issue Atlantic trailer Tan, moveable ramp, six metal wheels, crimpled axle, tan tow bar. Made 1955 — £10 £5 £3 £1

1a Various colours in first model, tan metal wheels crimpled axle — £10 £5 £3 £1

No 16 Second issue Atlantic trailer Tan, rigid ramp, eight plastic wheels, tow bar. Made 1957 — £10 £5 £3 £2

2a Tan with tan tow bar, eight grey plastic wheels, round axle — £7 £4 £3 £1

2b Orange with orange tow bar, eight grey plastic wheels, round axle — £7 £4 £3 £1

2c Orange with orange tow bar, eight black plastic wheels, knobby treads — £7 £4 £3 £1

2d Orange, black tow bar, eight black plastic wheels, knobby treads. Length 79 mm, 138-1 — £7 £4 £3 £1

No 16 Third issue Mountaineer dump truck with snow plough Grey cab and chassis, orange dump body, black plastic wheels, also grey plastic wheels on first model, orange and white stripes on front, length 76 mm, 109-1 TT scale — £7 £5 £3 £2

3a Red and white decal, closed step on ladder, hole in base, grey plastic wheels, fine treads — £5 £4 £2 £1

3b Red and white decal, closed step, hole in base, Black plastic wheels — £5 £4 £3 £1

3c Orange and white decal, closed step, hole in base, black plastic wheels — £5 £4 £3 £1

3d Orange and white decal, open step, no hole in base, black plastic wheels — £5 £4 £3 £1

THE GOLDEN YEARS OF LESNEY 17

15, Tractor truck. 16, Atlantic trailer

	Mint boxed	Mint unbxd	Good cond	Fair cond
1a Maroon, gold trim, no decals, metal wheels, crimpled axle, peaked roof	£10	£7	£5	£3
1b Maroon, gold or yellow trim, with decals, peaked roof, metal wheels, crimpled axle	£10	£7	£5	£3
1c Blue, silver trim, with decals, peaked roof, metal wheels, crimpled axle	£10	£7	£5	£3
1d Green, silver trim and decals, peaked roof, no window in rear of cab, metal wheels, crimpled axle	£7	£5	£3	£2
1e Green, silver trim with decals, peaked roof, window in rear of cab, metal wheels, crimpled axle	£7	£5	£3	£2
1f Green, silver trim, removal decal outline in black, peaked roof, metal wheels, crimpled axle	£7	£5	£3	£2
1g Green, silver trim, white decal without outline, 17 under cab, rounded roof, metal wheels, crimpled axle	£10	£7	£5	£3
1h Green, silver trim, decal outlined in black, 17 under cab, rounded roof, metal wheels, crimpled axle	£10	£7	£5	£3
1j Dark green, silver trim, decal outlined in black, rounded roof, grey plastic wheels, crimpled axle	£10	£7	£5	£3
1k Dark green, silver trim, decal outlined in black, rounded roof, plastic wheels, round axle	£6	£4	£2	£1

No 17 Second issue Austin taxi cab Red or maroon, silver grill, headlights and bumper, tan driver, light grey interior and base, grey plastic wheels. Made 1961

	Mint boxed	Mint unbxd	Good cond	Fair cond
	£15	£10	£5	£3
2a Deep maroon, grey plastic wheels	£10	£7	£5	£3
2b Deep maroon or bright red, silver plastic wheels	£10	£7	£5	£3
2c Deep maroon, silver plastic wheels	£10	£7	£5	£3

No 17 Third issue Hoveringham tipper Red cab and chassis, orange dumper, Hoveringham decal on tipper, black plastic wheels, length 76 mm 102-1, TT scale. Made 1964

	Mint boxed	Mint unbxd	Good cond	Fair cond
	£5	£4	£2	£1
3a Black half base, white plastic springs under the axles, a silver diamond in front	£5	£4	£3	£1

	Mint boxed	Mint unbxd	Good cond	Fair cond
3e Orange and white decal, open step, no hole in base, black plastic wheels	£5	£4	£3	£1
No 16 Fourth issue Case tractor bulldozer Red body with green tracks, yellow base, motor and blade, yellow plastic canopy. Made 1969	£4	£3	£2	£1
4a Black plastic rollers with no brace support at cab	£4	£3	£2	£1
4b Black plastic rollers with brace support at cab	£4	£3	£2	£1
4c Flaw: Case Tractor with no treads	£25	£10	£5	£3
No 16 Fifth issue Badger Red with silver trim, Rolls Royce front, white detector on roof, silver trim, six superfast black plastic wheels, length 70 mm, 75-1. Made 1974	75p	50p	25p	10p
5a Dark red with silver trim, white aerial on roof, length 68 mm. Made 1976				
No 17 First issue removal van Green or maroon, trim on radiator and bumper, Matchbox removals decal on sides, no number, 150:1. Made 1956	£10	£7	£5	£3

17, Removal van. 18, Caterpillar bulldozer

	Mint boxed	Mint unbxd	Good cond	Fair cond
3b Black half base, white plastic springs under the axle, red diamond in front	£5	£4	£2	£1
3c Red base, silver diamond in front, metal axle supports	£4	£3	£2	£1
3d Red base supports, red diamond in front, metal axle	£4	£3	£2	50p
3e Flaw: Missing decals, very rare, diamond half red, half silver	£100	£50	£15	£10

No 17 Fourth issue horse box Red cab, green box, grey doors, silver base and grill, black or blue plastic wheels, two horses, length 80 mm 92-1. Made 1969

	Mint boxed	Mint unbxd	Good cond	Fair cond
4a Ergomatic cab on flat base, rear tow slot, white or grey horses, black plastic wheels	£6	£4	£2	£1
4b Ergomatic cab on raised base, with or without reinforcement under door, black or blue plastic wheels	£4	£3	£2	£1
4c Mustard cabin, green box, superfast wheels	£4	£3	£2	£1
4d Red or orange cab, green box, superfast wheels	£5	£4	£2	£1
4e Red or orange cab, light or dark grey horsebox, white or grey horses, superfast wheels	£5	£4	£3	£2
Flaw: Horse box with no windows, very scarce	£100	£50	£10	£5

No 17 Fifth issue, the Londoner bus Red, with swinging London decals in blue, red and yellow, black lettering. First issue dated and numbered. Rare. Made 1973

	Mint boxed	Mint unbxd	Good cond	Fair cond
	£15	£10	£5	£3
5a Berger paints 'decals on side, paint brush at rear, superfast black plastic wheels	£6	£4	£3	£2
5b Paint brush at front, superfast black plastic wheels	£10	£6	£4	£2
5c Lighter bus body, swinging London decals, superfast wheels	£5	£4	£3	£2
Special issues: With Doncaster Impel 73 decals on the side of bus only a limited number issued	£50	£30	£10	£5
Flaw: Decal, swinging London first issue upside down	£25	£20	£7	£5
Flaw 2: No windows in top deck, only half on lower 78 mm	£250	£100	£10	£5
Doncaster Impel 76 in South Yorkshire colours	£5	£3	£2	£1

No 18 First issue caterpillar D8 bulldozer Yellow body and driver, red blade and side supports, metal rollers with rubber treads, no number on model, TT Scale 102-1. Made 1955

	Mint boxed	Mint unbxd	Good cond	Fair cond
	£7	£5	£3	£2
1a Yellow driver, metal rollers, crimpled axle	£5	£4	£2	£1
1b Yellow driver with black or blue hat, metal rollers, crimpled axle, no number	£5	£4	£2	£1

No 18 Second issue caterpillar bulldozer Yellow, yellow driver, tapered hood, metal rollers, rubber treads. Made 1959

	Mint boxed	Mint unbxd	Good cond	Fair cond
	£7	£5	£3	£2
2a 18 on back of blade, 8 under body, metal rollers, crimpled axle	£10	£7	£5	£3

THE GOLDEN YEARS OF LESNEY 19

	Mint boxed	Mint unbxd	Good cond	Fair cond
No 18 Third issue caterpillar bulldozer Yellow or orange, yellow driver, metal rollers, green rubber treads, length 61 mm, 102-1 TT scale. Made 1961	£6	£5	£3	£2
3a Number 8 under body, metal rollers, round axle	£5	£4	£3	£1
3b Number 18 on body, silver plastic rollers	£10	£5	£3	£2
3c Number 8 on body, black plastic rollers	£15	£10	£7	£5
3d Number 18 on body, black plastic rollers	£5	£4	£2	£1
No 18 Fourth issue caterpillar crawler bulldozer Yellow, no driver, green or black rubber treads, black plastic rollers, length 61 mm, 102-1. Made 1968	£5	£4	£3	£2
4a Small stack, colours as above	£5	£4	£3	£2
4b Small stack, hole in base	£4	£3	£2	£1
4c Large stack, hole in base	£5	£4	£2	£1
Flaw: no stack and half the engine missing, gives impression that it has been sheered off in casting	£250	£100	£25	£10
No 18 Fifth issue field car Yellow, tan plastic roof, white or bronze interior, spare tyre on rear, black plastic tyres, red hubs, silver trim, length 58 mm. Made 1969	£6	£5	£3	£2
5a Black base, grill, headlights, bumper, red or black hubs	£6	£5	£3	£2
5b Metal base, grill, headlights, bumper, red hubs	£6	£5	£3	£2
5c Metal base, green hubs	£6	£5	£3	£2
5d No number, superfast wheels, red, green or black hubs	£7	£5	£3	£1
5e Number, superfast wheels, red, green or black hubs	£7	£5	£3	£1
Flaw: Green body, black roof, half painted on one side, seems as though paint has run out whilst being finished	£150	£50		
Flaw: no wheels, extended hubs at front, no hubs at back, missing headlights	£100	£50		

	Mint boxed	Mint unbxd	Good cond	Fair cond
No 18 Sixth issue, Hondanora Red, silver and black motor cycle, red fin, decals in silver on black base, silver forks, handlebars, seven mudguards, length 64 mm, 35-1. Made 1975	£1	75p	50p	25p
6a as above but with special features added and black forks mudguards and handle bars	£2	£1	50p	25p
No 19 First issue MG sports car Tan, silver grill and headlights, tan driver, no number, spare tyre on trunk, red painted seats. Made 1955	£6	£5	£3	£2
1a White, metal wheels, crimpled axle	£6	£5	£3	£2
1b Cream, metal wheels, crimpled axle	£6	£5	£3	£2
No 19 Second issue MGA sports car Off white or cream, silver trim, tan driver, red base plate with number, red tail lights. Made 1958/59	£6	£5	£3	£2
2a Short steering post under steering wheel, metal wheels, crimpled axle	£6	£5	£3	£2

19 MG sports car. 20, State truck

THE GOLDEN YEARS OF LESNEY

	Mint boxed	Mint unbxd	Good cond	Fair cond
2b Short steering post, gold trim, grey plastic wheels, crimpled axle	£5	£4	£2	£1
2c No steering post, silver trim, grey plastic wheels, crimpled axle	£5	£4	£2	£1
2d Missing steering post, silver trim, grey plastic wheels, round axle	£4	£3	£2	£1
2e Missing steering post, silver trim, silver plastic wheels, round axle, OO scale	£4	£3	£2	£1
No 19 Third issue, Aston Martin racing car Green, silver grey driver, black numbered decal, 19 on rearsides, wire wheels, black tyres, length 64 mm, 61-1. Made 1962	£7	£5	£3	£2
3a Number 19 decal on rear	£5	£3	£2	£1
3b Number 52 decal	£5	£3	£2	£1
3c, 3d Number 5 and 41 decal, only a limited number produced	£15	£10	£5	£3
No 19 Fourth issue Aston Martin racing car Metallic green body, white driver, number 19 decals on rear, white wheels, black tyres, length 64mm, 61:1. Made 1964	£5	£4	£3	£2
4a Number 5, 7 and 41 decals all with this issue but very rare.	£100	£50		
No 19 Fifth issue Lotus racing car White driver, yellow hubs, yellow flashes and decals or stickers on sides and bonnet in various shades, black tyres, length 70 mm, 54-1. Made 1965	£5	£4	£3	£2
5a Orange body, three decals	£4	£3	£2	£1
5b Green body, three decals	£4	£3	£2	£1
5c Orange body, three sticker decals	£3	£2	£1	50p
5d Green body, three sticker decals	£3	£2	£1	50p
5e Metallic purple body, yellow sticker decals, superfast wheels	£5	£4	£3	£1
Flaw on 5e: all purple body, no decals, metallic one side only	£50	£10		

	Mint boxed	Mint unbxd	Good cond	Fair cond
No 19 Sixth issue road dragster Red body with silver motor, ivory interior, orange and black sticker decals on hood and trunk, length 76 mm, 65-1, superfast wheels. Made 1971	£2	£1	75p	50p
6a Number 8 sticker decals run vertically	£1	75p	50p	25p
6b Number 8 sticker decals run horizontally	£4	£3	£2	£1
Flaw on 6b: Only three wheels; the place for the fourth wheel is filled in at the front.	£25	£10		
No 19 Seventh issue road dragster Dark red or purple, metallic with white interior, silver engine and number 8 decals in yellow and orange circle, length 75 mm, 65-1. Made 1976				
No 20 First issue stake truck Dark red, silver grill, side tanks, headlights, ribbed bed, rear fenders, no number on model.	£10	£7	£5	£3
1a Dark maroon, metal wheels, crimpled axle	£7	£5	£3	£2
1b Light red or maroon, metal wheels, crimpled axle	£7	£5	£3	£2
1c Dark red, metal wheels, crimpled axle	£5	£4	£2	£1
1d Dark red, grey plastic wheels, crimpled axle	£5	£4	£2	£1
1e Dark red, grey plastic wheels, dark red side tanks, round axle	£5	£4	£2	£1
No 20 Second issue ERF 686 truck Blue, silver radiator, silver headlights, bumper with number, 'Everyday Ever Ready for life' decals on sides in blue, white and red trim, eight plastic wheels. Made 1959	£10	£7	£5	£3
2a Red body, silver radiator, headlights, bumper with number. 'Ever Ready, with life' decals on sides. Very rare model	£500	£100		
Flaw: Missing decals, red body	£1000	£250		
Flaw: Missing decals, blue body	£500	£150		
No 20 Third issue ERF 686 truck Blue with Ever Ready decals, blue or silver tank, silver radiator, grey plastic wheels, 125-1, crimpled axle. Made 1960	£10	£5	£3	£2

THE GOLDEN YEARS OF LESNEY 21

	Mint boxed	Mint unbxd	Good cond	Fair cond
3a Light blue body, grey plastic wheels, round axle	£6	£5	£3	£2

No 20 Third issue ERF 686 truck Low sided 8-wheel truck.

	Mint boxed	Mint unbxd	Good cond	Fair cond
3a Purple body, white decals 'Ever Ready for life' silver trim, grey plastic wheels, round or crimpled axles. Made 1961	£7	£5	£3	£2
3b Red and blue trim lines, length 70 mm, 125-1. Made 1962	£7	£5	£3	£2
3c Light grey plastic wheels, round or crimpled axle	£5	£4	£3	£1
3d Silver plastic wheels, knobby or fine treads, round axle	£5	£4	£3	£1
3e 8 wheeled open truck, orange decals, black plastic wheels, fine or knobby treads, last model issued. Made 1964	£7	£5	£3	£2
Flaw: One side only half painted, decals on one side	£100	£50		

No 20 Fourth issue taxi cab Yellow or orange, metal grill and base, taxi decals, length 76 mm, 72-1.00 scale. Made 1965

	Mint boxed	Mint unbxd	Good cond	Fair cond
	£7	£5	£3	£2
4a Orange yellow, grey base, grey plastic wheels, fine treads	£5	£3	£2	£1
4b Orange yellow, white or off white interior, grey base, black plastic wheels, fine treads	£5	£3	£2	£1
4c Orange yellow, metal base, taxi decal	£5	£4	£3	£2
4d Orange or yellow, red interior, metal or grey base, white or off white interior	£5	£4	£3	£2
4e Dark or bright yellow, red, off white, or white interior, light orange or yellow sticker decal. All models can have varied shades of decals	£4	£3	£2	£1

No 20 Fifth issue Lamborghini Marzel Metallic red, white or off white interior, amber windows, metal grill and base, superfast wheels, length 70mm, 58:1. Made 1970

	Mint boxed	Mint unbxd	Good cond	Fair cond
	£5	£4	£2	£1
5a Red hood with line imprint straight	£5	£4	£2	£1
5b Red hood with line imprint curved	£7	£5	£3	£2
5c Dark metallic red, orange or white interior. All models with superfast wheels. Made 1971	£5	£4	£3	£1
Flaw: Radiator grill missing, half painted one side	£100	£50		

No 20 Sixth issue Lamborghini Marzel Red, silver trim, white or off white interior. Made 1974

	Mint boxed	Mint unbxd	Good cond	Fair cond
	£5	£4	£2	£1
6a Two sticker decals, straight hood imprint, superfast wheels	£4	£3	£2	£1
6b Two sticker decals, curved hood imprint, superfast wheels	£4	£3	£2	£1
6c Salmon coloured or light yellow body, silver trim, curved hood, superfast wheels	£4	£3	£2	£1

No 20 Seventh issue police patrol car White with orange flash on side, blue light on roof, police decals in blue on each side, wide wheels, pin axles, length 82 mm, 61-1. Made 1975

	Mint boxed	Mint unbxd	Good cond	Fair cond
7a Orange light on roof. Made 1976		50p	25p	
Flaw: No flash, no decals, all white body	£10	£5		

No 21 First issue long distance coach Green, silver trim, orange lettered decal on red background, 'London to Glasgow', no number, metal wheels

	Mint boxed	Mint unbxd	Good cond	Fair cond
	£10	£7	£5	£3
1a Lighter green coach, darker red background on decals	£10	£7	£5	£3

21 Second issue long distance coach Green, silver trim, orange lettered decal, 'London to Glasgow' on red background, with number, 133-1. Made 1958

	Mint boxed	Mint unbxd	Good cond	Fair cond
	£7	£5	£3	£2
2a Light green, metal wheels	£7	£5	£3	£2
2b Much lighter green, grey plastic wheels, crimpled axle	£7	£5	£3	£2
2c Dark green, grey plastic wheels. Some models have crimpled, others round axles	£10	£5	£3	£2

No 21 Third issue milk delivery truck Light blue body, cream roof, grey seats, white bottles, metal wheels,

22 THE GOLDEN YEARS OF LESNEY

	Mint boxed	Mint unbxd	Good cond	Fair cond
silver trim, length 58 mm, 74-1, no decals on sides, no roof sign. Made 1961	£25	£15	£5	£3
3a Dark blue body, no decals on roof, white interior, white seats, silver trim	£250	£100		
No 21 Fourth issue milk delivery truck Light green, plastic windows, decals 'drink more milk' on top panel with number, black and red signs on doors, Made 1962	£7	£5	£3	£2
4a Cream bottles, clear windows, silver plastic wheels, knobby treads, bottle decals on doors, snap in base, clear decal on top	£5	£4	£3	£2
4b Green tinted windows, silver plastic wheels, cream bottles, knobby treads, clear decal	£5	£4	£3	£2
4c Cow decal, white decal on top, bottles,	£5	£4	£3	£2
4d Cow decal, white decal on top, white bottles, small grey plastic wheels	£7	£5	£3	£2
4e Cow decal, white decal on top, white bottles silver or black plastic wheels	£5	£4	£2	£1
4f Cow decals, white decal on top, white bottles, black plastic wheels, two rivets in base	£5	£4	£2	£1

	Mint boxed	Mint unbxd	Good cond	Fair cond
4g Cow decals, white decal on top, white bottles, black plastic wheels, two rivets in base, no silver trim on bumper	£4	£3	£2	£1
No 21 Fifth issue Foden concrete truck Yellow cab, orange yellow plastic barrel, red chassis, dark yellow tinted windows. Red filler at back on top, black plastic wheels, length 76 mm. Made 1969	£15	£10	£5	£3
No 21 Sixth issue Foden concrete truck Red cab and chassis, plastic barrel, green tinted windows, eight black plastic wheels. Made 1970	£10	£5	£3	£2
6a Orange yellow cab, metal rivet on end of barrel	£5	£4	£2	£1
6b No rivet on end of barrel	£5	£4	£2	£1
6c Yellow with light yellow barrel, red chassis, green base, cab base clips into front bumper, superfast wheels	£4	£3	£2	£1
6d Cab base not clipped to bumper, licence plate, superfast wheels	£5	£4	£2	£1
6e Light yellow body, superfast wheels	£4	£3	£2	£1
Flaw: part of chassis at rear with top scoop missing	£100	£50		
No 21 Seventh issue, rod roller Mustard body, red seat, red, blue and white decal on top at front, orange on outside rollers at rear, black plastic rollers, length 64mm, 77-1. Made 1973	50p	25p		
7a Flowered decal on top at front, black rollers, length 70 mm. Made 1974	50p	25p		
No 22 First issue sedan Black base, tow hook, silver trim, no number, no windows, metal wheels, crimpled axle. Made 1955	£10	£5	£3	£2
1a Red body, white roof, black base, small brace over front axle	£6	£5	£3	£2
1b Light maroon body, cream roof, black base, small brace over front axle	£6	£5	£3	£2
1c Dark maroon body, cream roof, black base, small brace over front axle, bad flaw in body where paint has missed one side	£50	£10		

21, Long distant coach. 22, Sedan

THE GOLDEN YEARS OF LESNEY

	Mint boxed	Mint unbxd	Good cond	Fair cond
1d Dark maroon body, cream roof, very dull black base	£5	£4	£3	£2
No 22 Second issue Vauxhall Cresta Silver grill, headlights and bumpers, red tail lights, black base with number and tow hook. Made 1958	£6	£5	£3	£2
2a Bright pink, metal wheels, crimpled axle	£100	£50	£10	£5
2b Pink, no windows, grey plastic wheels, crimpled axle	£100	£50		
2c Cream, no windows, grey plastic wheels, round axle	£50	£25		
2d Light pink, clear windows, grey plastic wheels, round axles	£5	£4	£3	£2
2e Two tone torquoise-bronze, grey plastic wheels, crimpled axle, clear windows	£6	£5	£3	£2
2f Two tone torquoise-bronze, clear windows, scale 74:1, grey plastic wheels, round axle	£5	£4	£3	£2
2g Two tone orchid grey, clear windows, grey plastic wheels, round axle	£5	£4	£3	£2
2h Two tone orchid grey, clear windows, silver plastic wheels, round axle	£4	£3	£2	£1
2j Light metallic gold, clear windows, rear gold bumper, silver plastic wheels, round axle	£5	£4	£3	£2
2k Deep dark metallic gold, tinted windows, gold rear bumper, silver plastic wheels, round axles	£4	£3	£2	£1
2l Dark metallic gold, tinted windows, silver rear bumper and tail lights, silver plastic wheels, round axles	£5	£4	£2	£1
Flaws: Very short nose on car, paint badly blistered, one colour of grey, one tone missing	£200	£50		
No 22 Third issue Vauxhall Cresta Two tone cream at top, blue underneath, silver trim, grey plastic wheels. 1958-64 series. Made 1961	£50	£25	£10	£5
3a Bronze, bronze rear bumper, tinted windows, grey plastic wheels, round axle	£5	£4	£3	£2
3b Bronze, bronze or silver bumper, black plastic wheels, knobby treads, round axle	£5	£4	£3	£2
3c Bronze, bronze or silver rear bumper, black plastic wheels, fine treads, round axle Tones of cars may vary	£5	£4	£3	£2
No 22 Fourth issue Pontiac Grand Prix Red, light grey interior, clear windows, silver grill and headlights, tow hook, black plastic wheels, length 76 mm, 72-1, 00 scale. Made 1965	£10	£7	£5	£3
4a Red or light red, no patent number, doors open, plastic spring	£5	£4	£3	£2
4b With patent number, doors open with metal spring	£5	£4	£3	£2
4c With patent number, doors open, metal spring, tow slot	£5	£4	£3	£2
4d Dark metallic purple or blue, superfast wheels, doors on this model do not open	£4	£3	£2	£1
4e Light purple, doors do not open, superfast wheels. Colours can vary for all above models	£4	£3	£2	£1
No 22 Fifth issue Freeman inter city commuter Pink or red body, off white interior, metal grill, base with tow slot, yellow decal on doors, length 76 mm, 55-1, wide superfast wheels	75p	50p	25p	15p
No 22 Sixth issue Blaze Buster Red, yellow plastic ladder, fire decals on rear sides, silver trim on hose and pipes, orange tinted windows, length 77 mm, 90-1. Made 1976				
No 23 First issue Berkeley Cavalier trailer Blue with decal 'On Tow' MBS 23 at lower right corner, tow bar, OO.HO scale, 80:1. Made 1956	£10	£5	£3	£2
1a Light blue, closed axle decal on corner, tow bar, metal wheels, crimpled axle, no number	£10	£5	£3	£2
1b Light blue, decals as above, metal wheels, crimpled axle open with number	£6	£5	£4	£2
1c Lime green with details as above, grey plastic or metal wheels, round axles on various models	£6	£4	£2	£1
Flaw: Lime green, no decals on tow bar, crimpled axle on one side only, fault in finishing	£50	£10		

23, Berkeley caravan. 24, Hydraulic excavator

	Mint boxed	Mint unbxd	Good cond	Fair cond
No 23 Second issue Bluebird caravan trailer Dark pink or fawn, decal lower right rear side with number, length 64 mm, 80:1, HO.OO scale, doors open. Made 1961	£10	£6	£4	£2
2a Pinkish tan, grey plastic wheels, round axle, flat tow bar	£7	£5	£3	£2
2b Tan, small grey plastic wheels, round axle, flat tow bar	£5	£4	£2	£1
2c Tan, flat tow bar, silver plastic wheels, knobby treads	£5	£4	£3	£1
2d Tan or pink, reinforced circle on tow bar, grey plastic wheels, knobby treads	£5	£3	£2	£1
2e Tan, reinforced circle on tow bar, grey plastic wheels, fine treads	£4	£3	£2	£1
Flaw: Paint missing, bad oversight by checker, very rare	£100	£50		

	Mint boxed	Mint unbxd	Good cond	Fair cond
No 23 Third issue trailer caravan Yellow body, red interior, white plastic removable roof, length 76 mm 72-1. OO scale, flat towbar, four black plastic wheels. Made 1965	£10	£7	£5	£3
3a Yellow, pale blue interior, removable roof, open axle, black plastic wheels, knobby treads	£5	£4	£3	£2
3b Yellow, blue interior, black plastic wheels, open axle, also model with closed axle, fine treads, opening roof	£5	£4	£3	£2
3c 3d 3e All three models have the following variations between them. Pink with closed axle, knobby treads and fine treads, opening roof	£5	£4	£3	£2
No 23 Fourth issue Volkswagen camper Blue with orange roof, light orange interior, clear windows, superfast wheels, length 67 mm, 63:1. Made 1970	£4	£3	£2	£1
4a and **4b** Very light or dark blue, otherwise details as above, decals at rear	£4	£3	£2	£1
No 23 Fifth issue Volkswagen camper Bright orange with slightly darker orange roof, decals on sides at rear, superfast wheels, length 67 mm, 63:1. Made 1974	£4	£3	£2	£1
No 23 Sixth issue Atlas wagon Purple cab and chassis, orange tipper, wide superfast wheels, off white interior of cab, clear windows, special action feature, length 71mm, 88:1. Made 1976				
No 24 First issue Weatherhill hydraulic excavator Orange, silver trim on muffler and hydraulic cylinders, small wheel in front, large wheels in rear, Lesney cast inside scoop with number. Made 1956	£10	£7	£5	£3
1a Orange and yellow, silver trim, metal wheels, crimpled axle	£7	£5	£3	£2
1b Orange, metal wheels, crimpled axle, decal at rear 'Weatherhill Hydraulic' also in yellow	£7	£5	£3	£2
No 24 Second issue Weatherhill hydraulic excavator Grey plastic wheels, crimpled axle and no trim, Lesney removed from scoop, cast inside body, OO scale, 75:1	£5	£4	£2	£1

THE GOLDEN YEARS OF LESNEY 25

	Mint boxed	Mint unbxd	Good cond	Fair cond
2a Grey plastic wheels, round axle, thick driver	£5	£4	£2	£1
2b Black plastic wheels, round axle, fine tread on front wheel, light imprint 'Lesney, England', length 67 mm	£5	£4	£2	£1
Flaw: missing driver on orange and yellow models, factory fault	£50	£25		

No 24 Third issue Rolls Royce Silver Shadow Blue metallic, off white interior, metal grill and base, silver plastic hubs, black plastic tyres, length 76 mm, 67:1, Made 1967

	£7	£5	£3	£2
3a No patent number, cast in front base	£7	£5	£3	£2
3b Metallic red with patent number, tow slot and small tab, also in darker shades of red. Made 1967	£10	£5	£3	£2
3c Light metallic red, black base, superfast wheels, opening boot	£7	£5	£3	£2
3d Darker red with black base, superfast wheels	£7	£5	£3	£2
3e Grey base, wide wheels, superfast wheels	£5	£4	£3	£2
3f Light metallic green, wide wheels, very rare model	£250	£100		
3g Metallic pink, pink base, wide superfast wheels	£25	£15	£10	£5

No 24 Fourth issue team Matchbox car Bright yellow, white driver, silver plastic engine in rear, blue decal on front with red outline and the number 8 in black letters on a white circle, Matchbox Team decals in yellow, wide plastic wheels, length 76mm, 54:1. Made 1973

	£7	£5	£3	£2

No 24 Fifth issue team Matchbox racer Metallic red body, with decals as fourth issue, wide black plastic wheels, length 73mm, 54-1. Made 1974

	£1	75p	50p	25p

5a Deep red metallic, white driver, otherwise details as above, length 75mm. Made 1976

	£1	75p	50p	25p

No 25 First issue Dunlop truck Bedford 12cwt van Blue, silver radiator, headlights and bumpers, black base, metal wheels, Dunlop decals. Made 1956

	£15	£10	£5	£3

	Mint boxed	Mint unbxd	Good cond	Fair cond
1a Metals wheels, crimpled axle, orange background with letters Dunlop	£10	£5	£3	£2
1b Grey plastic wheels, crimpled axle, orange background with black letters, also light yellow background with black letters	£7	£5	£3	£2
1c Light yellow or orange background with black letters, grey plastic wheels, round axle	£7	£5	£3	£2
1d Light yellow background, black letters, small grey plastic wheels, round axle	£7	£5	£3	£2
Flaw: Van with grey plastic wheels, Dunlop decals on one side missing	£50	£25		

No 25 Second issue Volkswagen sedan Silver blue, silver headlights, grey plastic wheels, plastic windows, opening in boot, length 62mm, OO scale, 65:1. Made 1961

	£7	£5	£3	£2

25, Bedford Dunlop truck. 26, Concrete truck

26 THE GOLDEN YEARS OF LESNEY

	Mint boxed	Mint unbxd	Good cond	Fair cond

2a Clear windows, red tail lights, silver exhaust louvres on motor cover, grey plastic wheels, knobby treads — £7 £5 £3 £2

2b Clear windows, blue tail lights and exhaust, grey plastic wheels, knobby treads — £7 £5 £3 £2

2c and **2d** Tinted windows, light grey plastic wheels, knobby treads, silver plastic wheels, knobby treads — £5 £4 £3 £2

2e Tinted windows, silver plastic wheels, knobby treads, blue rear bumper — £5 £4 £3 £2

2f Tinted windows, black plastic wheels, knobby treads — £4 £3 £2 £1

No 25 Third issue BP tanker Orange hinged cab, white tank, green chassis, yellow 'BP' inside green shell decal, opening cab, six grey plastic wheels, length 76 mm, 85:1, HO scale. Made 1965 — £10 £7 £5 £3

3a Yellow hinged cab, grey plastic wheels, round axle, also model with black plastic wheels, round axle — £7 £5 £3 £2

3b Blue cab, white tank, blue chassis, 'Aral' decal, only a few made, very rare model — £50 £25 £15 £10

No 25 Fourth issue Ford Cortina GT Light blue body, silver metal grill and base, clear windows, off white interior, tow hook, black plastic wheels, length 67 mm, 62:1. Made 1967 — £7 £5 £3 £2

No 25 Fifth issue Ford Cortina GT Light brown, silver pink radiator, black plastic wheels, white interior, tow box. Made 1969 — £5 £4 £3 £2

5a Light brown with tow box, yellow plastic roof rack — £5 £4 £3 £2

5b Light brown, superfast wheels, — £4 £3 £2 £1

5c Metallic blue, superfast wheels. Above models are in various shades — £4 £3 £2 £1

No 25 Sixth issue mod tractor Pink, silver trimmed engine, large wide plastic wheels at rear, small wide wheels at front, yellow seat, length 57 mm, 52:1. Made 1973 — 50p 40p 15p 10p

---

	Mint boxed	Mint unbxd	Good cond	Fair cond

6a Light pink, silver plastic hubs — 50p 40p 15p 10p

6b Dark purple body. Made 1976

No 26 First issue concrete truck Orange, four wheels, eight paddles inside body barrel. Made 1956 — £6 £5 £3 £2

1a Gold trim on radiator, headlights and side tanks, short stem on barrel, metal wheels, crimpled axle — £5 £3 £2 £1

1b Silver trim, short stem on barrel, metal wheels, crimpled axle — £5 £3 £2 £1

1c Silver trim, long stem on barrel, metal wheels, crimpled axle — £5 £3 £2 £1

1d and **1e** Silver radiator and headlights, grey plastic wheels, crimpled axle, or grey plastic wheels, round axle — £5 £3 £2 £1

1f Silver radiator and headlights, silver plastic wheels, round axle, length 4.4 cms, 123:1, TT scale — £4 £3 £2 £1

No 26 Second issue Readymix concrete truck Orange chassis and body, grey barrel, grey wheels, length 64mm, 98-1, HO scale. Made 1962 — £7 £5 £3 £2

27, Bedford low loader. **28,** Bedford compressor truck

THE GOLDEN YEARS OF LESNEY 27

	Mint boxed	Mint unbxd	Good cond	Fair cond
2a Orange chassis, orange body, grey plastic wheels, length 64mm	£5	£4	£3	£2
No 26 Third issue Readymix concrete truck Orange barrel, grey plastic wheels, knobby treads, silver trim, small brace on base behind side gas tanks. Made 1963	£5	£4	£3	£2
3a Orange barrel, grey plastic wheels, fine treads	£4	£3	£2	£1
3b Orange barrel, grey plastic wheels, fine treads, silver trim	£4	£3	£2	£1
3c Orange barrel, black plastic wheels, fine treads, silver trim	£4	£3	£2	£1
3d Orange barrel, black plastic wheels, fine treads, no silver trim	£4	£3	£2	£1
No 26 Fourth issue GMC tipper truck Red tipping cab, red chassis, yellow body, green tinted windows, length 67 mm, 86:1. Made 1968	£5	£4	£3	£2
4a Red tipping cab, green chassis, silver metal body, green tinted windows, black plastic wheels	£5	£4	£3	£2
4b Black plastic base, superfast wheels	£3	£2	£1	75p
4c Black plastic base, superfast wheels, reversed reading, all models come with various die cast numbers on body and chassis	£3	£2	£1	75p
Flaw: 3-4 spoked wheels, 1-8 spoked	£50	£25		
No 26 Fifth issue, Big Banger Red body with silver engine through bonnet, blue tinted windows, big banger decals on both sides, silver exhausts at each side, wide black plastic wheels, length 76mm, 64:1. Made 1973	£1	50p	25p	10p
5a Lighter pink, details as above. Made 1974	50p	40p	20p	10p
5b Very dark body, otherwise details same. Made 1976				
No 27 First issue Bedford low loader Silver radiator, headlights and bumper, gas tank on side, four wheels on tractor, two wheels on trailer. Made 1956	£10	£7	£5	£3
1a Green cab, tan trailer, metal wheels, crimpled axle	£10	£7	£5	£3

	Mint boxed	Mint unbxd	Good cond	Fair cond
No 27 Second issue Bedford low loader Silver radiator, bumper and headlights, four wheels on trailer, 133:1. Made 1960	£7	£5	£3	£2
2a Green cab, light tan trailer, grey plastic wheels, crimpled axle, green bumper	£7	£5	£3	£2
2b Dark green cab, light tan trailer, grey plastic wheels, green bumper, round axle	£7	£5	£3	£2
No 27 Third issue Cadillac sedan Purple body, silver grill, headlights and bumpers, tan roof, red tail lights and tow hook, length 69mm, 80:1, OO scale. Made 1961	£7	£5	£3	£2
3a Green body, white roof, clear windows, silver plastic wheels, knobby treads, red base	£10	£5	£3	£2
3b Silver grey body, cream roof, clear windows, red base, silver plastic wheels, knobby treads	£5	£4	£3	£2
3c Silver grey body, pink roof, clear windows	£5	£4	£3	£2
3d Lilac body, tannish pink roof, clear windows	£5	£4	£3	£2
3e Silver grey body, tannish pink roof, tinted windows, red base, rear bumper not silver, silver plastic wheels, knobby treads	£5	£4	£3	£2
3f Lilac body, tannish pink roof, details as 3e	£5	£4	£3	£2
3g Silver grey body, tannish pink roof, black base, other details as on 3e	£5	£4	£3	£2
3h Lilac body, tannish pink roof, grey plastic wheels, knobby treads, black base, rear bumper unpainted	£15	£10	£5	£3
3j Lilac body, tannish pink roof, tinted windows, grey plastic wheels, knobby treads, black base, no paint on rear bumper or tail-lights	£15	£10	£5	£3
3k Lilac body, pink roof, tinted windows, black base, no paint on bumper or tail light, grey plastic wheels, fine treads	£15	£10	£5	£3
3l Lilac body, tannish pink roof, tinted windows, grey plastic wheels, knobby treads on front, fine treads on rear, black base, no paint on rear bumper and tail lights	£25	£10	£7	£5

28 THE GOLDEN YEARS OF LESNEY

	Mint boxed	Mint unbxd	Good cond	Fair cond
3m Lilac body, pink roof, black plastic wheels, knobby treads on front, black plastic wheels, fine treads on rear	£10	£5	£3	£2
No 27 Fourth issue Cadillac sedan Pink body, tan roof, also done in green and yellow, silver trim, rear bumper and tail lights, streamlined model with sharp fins at rear, length 70mm, 80:1, scale OO. Made 1965	£6	£5	£3	£2
Flaw: Paint missing from roof, missing fin	£30	£15	£10	£5
No 27 Fifth issue Mercedes 230 SL Convertable in white, silver metal grill and base, red seats, clear plastic windshield, red tow hook with pattern BPW 1966. Length 71mm, 60-1. Made 1966	£7	£5	£3	£2
5a Cream, plastic door hinge	£5	£4	£2	£1
5b White, plastic door hinge	£5	£4	£2	£1
5c Cream, metal door hinge, superfast wheels	£6	£4	£2	£1
5d White, metal door hinge, superfast wheels	£6	£4	£2	£1
5e Light or dark yellow, metal door hinge, superfast wheels	£6	£5	£3	£2
5f Light or dark yellow, black seats and tow hook, superfast wheels	£10	£7	£5	£3
5g Blue body, black seats, clear plastic windshield, black plastic superfast wheels, tow hook, rare model	£25	£10	£5	£3
No 27 Sixth issue Lamborghini Countach Bright orange body, red interior, '9' decal on nose; opening in rear to reveal engine, length 76mm, 54:1. Made 1974	£1	75p	50p	25p
6a Dark mustard with 'nine' decal	50p	40p	25p	15p
6b Pink, yellow flashes on top and on hatch that covers engine, black flashes on rear, Number '8' decal in red on black square with red border outline, length 74mm. Made 1976				
No 28 First issue, Bedford compressor truck Orange, silver grill on front and rear, silver headlights and bumper, silver hubs on metal wheels, orange tanks behind cab, six wheels, OO scale. Made 1956	£7	£5	£3	£2
1a Dull orange body, metal wheels, crimpled axle	£5	£4	£3	£2
1b Yellow body, metal wheels, crimpled axle	£5	£4	£3	£2
No 28 Second issue Thames trader compressor truck Silver grills front and rear, front bumper and hood grill, rivet front base, spread rivet rear base, OO scale, 75:1. Made 1960	£10	£7	£5	£3
2a Yellow orange, black plastic wheels, crimpled axle, knobby treads, middle front grill silver	£5	£4	£3	£2
2b Yellow orange, black plastic wheels, round axle, knobby treads, full front grill silver, rear grill no silver	£5	£4	£3	£2
2c Yellow orange, black plastic wheels, round axle, full front grill silver, rear grill with full silver	£10	£5	£3	£2
2d Yellow, black plastic wheels, round axle, knobby treads, middle front grill silver, rear grill silver	£5	£4	£3	£2
2e Yellow, black plastic wheels, round axle, knobby treads, middle front grill, rear all silver	£6	£5	£4	£3
2f Light yellow, round axle, knobby treads, full front and rear grills silver. This model has also some wheels with fine treads	£5	£4	£3	£2
Flaw: Details as on previous model except front wheels have fine treads and rear wheels have knobby treads, tanks behind cab missing top part	£100	£50		
No 28 Third issue Mark 10 Jaguar Light blue, off white interior, tow hook, hood opens, black base, black plastic wheels, silver trim, length 70mm, 72:1, OO scale. Made 1964	£10	£5	£3	£2
3a Light brown, yellow interior, tow hook where back opens, hood opens, black base, black plastic wheels. Made 1965	£100	£50	£15	£10
3b Light brown, off white interior, tow hook, hood opens, black base, black plastic wheels. Made 1965	£7	£5	£3	£2
3c Brown motor, silver bumper, black licence plate	£7	£5	£3	£2

THE GOLDEN YEARS OF LESNEY

	Mint boxed	Mint unbxd	Good cond	Fair cond
3d Brown motor, silver bumper and licence plate	£7	£5	£3	£2
3e Silver metal motor, silver bumper and licence plate	£5	£4	£3	£2
3f Silver metal motor; silver bumper, black licence plate	£5	£4	£3	£2
3g Dark fawn, blue interior, silver bumper and licence plate, silver metal motor. Special issue 1966	£25	£20	£15	£10
3h Light metallic brown body, white interior, silver bumper and number plate, light brown engine. Made 1968	£5	£4	£3	£2

No 28 Fourth issue Mack dump truck Reddish orange with metal grill, bumper and base, three step ladder to cab, green tinted windows, pattern overhang roof on dump body, orange hubs, tilting body, black knobby treads, length 67 mm, 118:1. Made 1969

4a Dark tan chassis and tipper, silver hubs, superfast wheels. Made 1970	£10	£7	£5	£3
4b Yellow or red plastic hubs	£7	£5	£3	£2
4c Yellow plastic hubs, no windows, part of cab windscreen very warped at one side	£7	£5	£3	£2
4d Pea green body, green windows, superfast wheels on balloon tyres, three steps open	£25	£10		
4e Pea green, steps now closed in, superfast wheels	£5	£4	£3	£2
4f Mustard body, chassis and tipper, silver grill and bumper, three open steps	£10	£7	£5	£3
	£5	£4	£3	£2

No 28 Fifth issue stoat Dark and light tan, army combat or space module vehicle, observation sentry in centre, wide superfast wheels, length 67mm, 52:1. Made 1974

	£1	75p	50p	25p
5a Light fawn with details as above, figure revolves. Made 1975	£1	75p	50p	25p

No 29 First issue Bedford milk truck Tan body, white bottle load, white base, silver grill, bumper and head lights, metal wheels, crimpled axle. OO scale, 75:1. Made 1956

	Mint boxed	Mint unbxd	Good cond	Fair cond
	£10	£7	£5	£3
1a Grey plastic wheels, crimpled axle, otherwise details as number 1	£6	£5	£3	£2
1b Grey plastic wheels, round axle, otherwise details as number 1	£5	£4	£3	£2

No 29 Second issue Austin Cambridge Two tone green, light on top and dark below, silver grill, bumper and headlights, red tail lights, windows, black base and tow hook, 70:1, scale OO, length 57 mm. Made 1961

	£10	£7	£5	£3
2a Clear windows, grey plastic wheels, round axle, knobby treads	£6	£5	£2	£1
2b Tinted windows; grey plastic wheels, round axle, knobby treads	£5	£3	£2	£1
2c Clear windows; silver plastic wheels, knobby treads	£5	£3	£2	£1
2d Clear or tinted windows, silver plastic wheels, knobby treads, green rear bumper	£5	£3	£2	£1
2e Darker shades of green, yellow line trim. Made 1964	£5	£3	£2	£1
2f Dark blue, light blue roof, silver trim, only a few of these were put out for a special order, a worthwhile find. Made 1965	£50	£25		
2g Two tone green, tinted windows, green rear bumper, black plastic wheels, knobby treads or fine treads on tyres	£5	£3	£2	£1
2h Tinted windows, green rear bumper, tail lights, black plastic wheels, fine treads	£10	£5	£3	£2

No 29 Third issue fire pumper Red, metal grill, bumper and base, blue tinted windows, white plastic hose and ladder on sides, 'Denva' decals on doors, blue light on roof of cab, four black plastic wheels, length 76 mm, 104:1, TT scale. Made 1966

	£7	£5	£3	£2
3a Orange and yellow body, with 'Denver' decals	£6	£5	£3	£2
3b Without decal	£5	£4	£2	£1

THE GOLDEN YEARS OF LESNEY

	Mint boxed	Mint unbxd	Good cond	Fair cond
3c Without decal and tow slot	£5	£4	£2	£1
3d With tow slot and without decal, raised panel on doors	£5	£4	£2	£1
No 29 Fourth issue fire pumper Light on cab, white ladder, white hose and accessories; raised panel on doors, superfast wheels, silver hubs. Made 1970	£5	£3	£2	£1
No 29 Fifth issue racing mini Red body, clear windows, white interior, metal grill and base, number 29 in black letters on yellow background decals on doors, superfast wheels, length 57mm, 53:1. Made 1971	£2	£1	50p	25p
5a Metallic bronze decals, superfast wheels	£1	75p	50p	25p
5b Flaw: Metallic bronze with no decals	£25	£10		
5c Metallic bronze, sticker decals in yellow with light orange border, black number 29	£1	75p	50p	25p
5d Dark metallic bronze, sticker decals as 5c	£1	75p	50p	25p
No 29 Sixth issue racing mini Metallic red, also metallic pink, decals in yellow and black letters, grey interior, wide superfast wheels. Made 1976				
No 30 First issue Ford prefect Grey, silver grill, headlights and bumpers, red tail lights, black base with tow hook, OO scale. Made 1956	£10	£5	£3	£2
1a Tan, metal wheels, crimpled axle	£7	£5	£3	£2
1b Tan, grey plastic wheels, crimpled or round axles	£7	£5	£3	£2
1c Light blue colour, metal wheels, crimpled axle	£6	£5	£2	£1
1d Light blue, grey plastic wheels, crimpled or round axle	£5	£4	£2	£1
No 30 Second issue Ford prefect Yellow green, silver headlights, red tail lights, tow hook, metal or grey plastic wheels, OO scale, 71:1, rare model. Made 1960	£50	£15	£10	£5
No 30 Third issue 6 wheel crane truck Silver grey, orange crane and hook, black base under cab, metal wheels, length 67 mm, 128:1. Made 1961	£10	£7	£5	£3

	Mint boxed	Mint unbxd	Good cond	Fair cond
3a Light orange crane with open bottom, orange metal hook with ball, grey plastic wheels, round axle, knobby treads	£5	£4	£2	£1
3b Light orange crane with open bottom, silver metal hook with ball, silver plastic wheels, round axle, knobby treads	£5	£4	£2	£1
3c Dark orange crane with open bottom, silver metal hook with ball, silver plastic wheels, round axle knobby treads	£5	£4	£2	£1
3d Light or dark orange crane with closed bottom, silver metal hook with ball, black plastic wheels, knobby treads	£7	£5	£3	£2
3e Light or dark orange crane with closed bottom; silver metal hook with triangle; black plastic wheels, knobby treads, round axle	£6	£5	£3	£2
3f Light orange or dark orange crane with closed bottom, grey plastic hook, black plastic wheels, knobby treads, round axle	£5	£4	£2	£1
No 30 Third issue 6 wheel crane truck Silver grey, orange crane and hook, black base under cab, metal wheels, length 67 mm, 128:1. Made 1961	£10	£7	£5	£3
3a Light orange crane with open bottom, orange metal hook with ball, grey plastic wheels, round axle, knobby treads	£5	£4	£2	£1
3b Light orange crane with open bottom, silver metal hook with ball, silver plastic wheels, round axle, knobby treads	£5	£4	£2	£1
3c Dark orange crane with open bottom, silver metal hook with ball, silver plastic wheels, round axle, knobby treads	£5	£4	£2	£1
3d Light or dark orange crane with closed bottom, silver metal hook with ball, black plastic wheels, knobby treads	£7	£5	£3	£2
3e Light or dark orange crane with closed bottom, silver metal hook with triangle, black plastic wheels, round axle, knobby treads	£6	£5	£3	£2

THE GOLDEN YEARS OF LESNEY

	Mint boxed	Mint unbxd	Good cond	Fair cond
3f Light or dark orange crane with closed bottom, grey plastic hook, black plastic wheels, round axle, knobby treads	£5	£4	£2	£1
3g Light or dark orange crane with closed bottom, grey plastic hook, black plastic wheels, round axle, fine treads	£4	£3	£2	£1
No 30 Fourth issue 6 wheel crane truck Silver bronze cab and chassis, orange crane, silver bronze wheels, hook, length 67mm, 128-1. Made 1963	£10	£7	£5	£3
4a Green metallic silver body, orange crane, and green metallic hook. Made 1964	£10	£7	£5	£3
4b Fine bronze metallic body, orange crane, bronze hook. Made 1965	£10	£7	£5	£3
No 30 Fifth issue 8 wheel crane truck Green cab and chassis, orange crane, yellow plastic hook, black plastic wheels, metal half base, length 76mm, 114-1, TT scale. Made 1966	£10	£7	£5	£3
5a Details as above apart from red plastic hook	£10	£7	£5	£3
5b Details as above apart from yellow plastic hook, and tow slot	£10	£7	£5	£3
No 30 Sixth issue 8 wheeled crane Red body, golden metallic crane, golden hook, superfast wheels, silver trim, length 76mm, 114-1. Made 1970	£5	£4	£2	£1
6a Red body, orange crane, superfast wheels, silver trim	£5	£4	£2	£1
No 30 Seventh issue beach buggy Pink with gold paint spots, cream interior, dark orange side tanks, large superfast wheels, length 51mm, 53-1.	£1	75p	50p	25p
7a Large wheels with thick or thin treads	£1	75p	40p	10p
7b White interior with side tanks, thin treads	£1	75p	40p	10p
7c Bright yellow side tanks, pink body with yellow spots	75p	50p	25p	10p
7d Dark red body, side tanks, wide superfast wheels. Made in 1976				

	Mint boxed	Mint unbxd	Good cond	Fair cond
No 31 First issue American Ford station wagon Yellow, single silver headlights, bumpers and hood ornament, red tail lights with black base, metal wheels, crimpled axle, OO scale, 75:1. Made 1956	£10	£5	£3	£2
1a Grey plastic wheels, crimpled axle, other details as above	£5	£3	£2	£1
1b Grey plastic wheels, round axle, other details as above	£5	£3	£2	£1
No 31 Second issue American Ford station wagon Metallic green body with peach roof, silver headlights and bumpers, red tail light, tinted windows, length 69mm, 75:1, OO scale. Made 1961	£10	£5	£3	£2
2a Light metallic green, pink roof, green base, silver plastic wheels, knobby treads	£7	£5	£3	£2
2b Metallic green, pink roof, green rear bumper, silver trim, silver plastic wheels, very fine treads	£7	£5	£3	£2
2c Very dark metallic green, peach roof, silver plastic wheels, knobby treads, no trim	£5	£4	£2	£1
No 31 Third issue American station wagon Dark green body, non metallic, pink roof, red tail lights, silver trim, grey plastic wheels, black base, very fine treads on wheels, length 70mm 75:1, OO scale. Made 1963	£10	£7	£5	£3
3a Metallic green, pink roof, red tail lights, red base, grey plastic wheels, very fine treads. Made 1964	£5	£3	£2	£1
No 31 Fourth issue Lincoln Continental Royal blue, black trim, silver grey interior, opening boot, silver bumper at rear and front, black metal wheels, length 76mm, 74:1, OO scale. Made 1965	£25	£15	£7	£5
No 31 Fifth issue Lincoln Continental Sky blue, off white interior, clear windows, opening trunk, metal grill, bumpers and base, black plastic wheels, length 76mm, 74-1, OO scale. Made 1966	£10	£5	£3	£2
5a Medium blue, black plastic wheels	£4	£3	£2	£1
5b Dark blue with tow slot	£4	£3	£2	£1

THE GOLDEN YEARS OF LESNEY

	Mint boxed	Mint unbxd	Good cond	Fair cond
5c Mint green with tow slots	£5	£3	£2	£1
5d Lime green, superfast wheels	£4	£2	£1	50p
5e Light lime green, superfast wheels	£4	£2	£1	50p
Flaw: Very rare find, only one known in existence, paint completely missing from body	£500	£100		
No 31 Sixth issue Lincoln continental Metallic gold body, white interior, opening boot, silver grill and bumpers, wide superfast wheels. Made 1971	£2	£1	50p	25p
No 31 Seventh issue Volks-dragon Light red body, decal eyes in black, orange and yellow on bonnet, silver mountings appearing from rear window, silver plastic headlights and bumpers. Made 1972	£2	£1	50p	25p
7a Pale pink with white mounting at back, eye decals on front. Made 1974	£1	75p	50p	25p
7b Metallic red, details as above. Made 1975	£1	75p	50p	25p
7c Dark red, details as 7b. Made 1976				
No 32 First issue Jaguar XK140 Off white, Jag coupe, Black base, silver grill, headlights and bumpers, red rear tail lights, front fender lights silver, crimped axle. Made 1956.	£10	£7	£5	£3
1a Off white, metal wheels, crimpled axles, front fender lights without paint	£7	£5	£3	£2
1b Flaw: Off white, grey plastic wheels, no paint on fender lights. Paint almost non existent on roof, OO scale	£100	£25	£10	£5
1c Red body, grey plastic wheels, round axle, length 60mm	£10	£5	£3	£2
1d Red body, red rear bumper, grey plastic wheels, round axle, OO scale, 75:1	£7	£5	£3	£1
No 32 Second issue E-type Jaguar Pale pink, black base, clear windows, pink interior, wire wheels, length 67 mm, 65:1. Rare model. Made 1963	£25	£10	£5	£3
2a Dark metallic red, green tinted windows, thin grey tyres	£10	£7	£5	£2

	Mint boxed	Mint unbxd	Good cond	Fair cond
2b Dark metallic red, clear windows, thin grey tyres	£5	£4	£3	£2
2c Dark metallic red, clear windows, thin black plastic tyres	£5	£4	£3	£2
2d Light metallic red, clear windows, thin black plastic tyres	£5	£4	£3	£2
2e Dark metallic red, clear windows, thick black plastic tyres	£5	£4	£3	£2
2f Light metallic red, clear windows, thick black plastic tyres	£4	£3	£2	£1
Flaw: Red paint missing almost completely from one side	£100	£50		
No 32 Third issue Leyland petrol tanker Dull green cab and chassis, white tank, Ergomatic Cab on base, BP decals on sides, black plastic wheels, length 76 mm, 92-1. Made 1968	£10	£7	£5	£3
3a Green cab with brighter green chassis, white tank, Ergomatic cab on base, BP decals on sides, eight black plastic wheels	£5	£4	£3	£2
3b Bright green cab and chassis, Ergomatic Cab on flat silver base, large BP sticker decals	£5	£4	£2	£1
3c Ergomatic Cab on raised silver platform, very large BP sticker decals	£4	£3	£2	£1
3d Ergomatic cab on raised silver platform, tow slot, very large BP decals	£5	£4	£3	£2
3e Ergomatic Cab on flat white plastic base, tow slot, small BP sticker decals	£5	£4	£2	£1
No 32 Fourth issue Leyland petrol tanker Bright green cab and green chassis, Ergomatic cab with raised silver platform, tow slot with large or small BP sticker decals, black plastic wheels, length 76mm, 92:1. Made 1969.	£10	£5	£3	£2
4a Dark green Ergomatic cab, other details as above	£6	£5	£3	£1
4b Blue cab and chassis, white tank with Blue Aral sticker decals on sides and rear of tank, on raised silver base with or without tow slot. Special issue	£50	£15	£7	£5

THE GOLDEN YEARS OF LESNEY

	Mint boxed	Mint unbxd	Good cond	Fair cond
4c Green Ergomatic cab, superfast wheels, details as 4a. Made 1970	£5	£4	£3	£1
4d Blue cab and chassis, white tank with Blue Aral sticker decals on sides and rear of tank, flat silver base with tow slot, superfast wheels. Made 1971	£25	£15	£7	£5
4e Green cab and chassis, white tank, small BP sticker decals on side, Ergomatic Cab on flat silver base, superfast wheels. Made 1971-72.	£5	£4	£3	£1
4f Ergomatic cab on flat grey base, tow slot, small BP sticker decals, superfast wheels. Last model made in 1972	£6	£5	£3	£2
No 32 Fifth issue Masserati Bora Pink, yellow interior, number 8 decal on bonnet with green line on a yellow background, opening doors, wide superfast wheels. Made 1973	£3	£2	£1	50p
5a Dark metallic body, yellow or orange interior, opening doors, no sticker decal on bonnet top, length 76mm. Made 1975	50p	30p	20p	10p
5b Very dark colour, other details as above. Made 1976				
No 33 First issue Ford Zodiac sedan Light blue, black base, silver grill, headlights, bumpers, red tail lights, tow hook. Made 1956	£10	£5	£3	£2
1a Loden green, metal wheels, crimpled axle	£7	£5	£3	£2
1b Dark blue-green, metal wheels, crimpled axles	£7	£5	£3	£1
1c Light blue-green, metal wheels, crimpled axles	£6	£4	£3	£1
1d Green, crimpled or round axle	£5	£4	£2	£1
1e Orange body, tan top, large grey plastic wheels, large or small knobby treads	£5	£4	£2	£1
No 33 Second issue Ford Zodiac sedan Orange body, tan top, clear plastic windows, small grey wheels, knobby treads. Made 1958	£7	£5	£3	£2
2a Orange body, grey top, clear plastic windows, small grey plastic wheels, knobby treads	£7	£5	£3	£2
2b Orange body, silver grey top, tinted windows, silver plastic wheels, knobby treads	£5	£4	£3	£2

	Mint boxed	Mint unbxd	Good cond	Fair cond
2c Orange body, tan top, tinted windows, silver plastic wheels, knobby treads	£5	£4	£3	£2
Flaw: No paint on body and only slight colour on roof, a rare find. Factory oversight	£100	£50		
No 33 Third issue Ford Zephyr Zodiac Very dark blue-green, clear windows, black base, fine silver line trim on first model, silver grill, headlights and bumpers, tow hook, grey plastic wheels, round axle, OO scale, 71:1. Made 1960	£25	£20	£10	£5
3a Medium dark blue, details as above	£10	£7	£3	£2
3b Very light blue body, grey plastic wheels	£7	£5	£3	£2
3c Dark, light and medium blue body, all with silver plastic wheels	£5	£4	£3	£2
3d Dark, light and medium blue body, black plastic wheels, with or without silver bumpers	£5	£4	£3	£2
No 33 Fourth issue Ford Zodiac Two tone light pink or off white on roof, deep sandy orange for lower body, silver trim on bumpers, headlights and grill, grey plastic wheels, length 60mm, 71:1, OO scale. Made 1961-62	£10	£7	£5	£3
4a Light, dark and medium blue, with or without silver trim, plain colours for bumpers etc, grey or black plastic wheels, length 64mm, OO scale. Made 1961-62.	£5	£4	£3	£1
Flaw: Grill and rear bumper missing	£50	£10		
No 33 Fifth issue Lamborghini Miura Golden yellow, silver trim, light fawn interior, black plastic wheels, length 70mm, 62:1. Made 1969	£5	£4	£3	£2
5a Yellow body, cream interior, with or without silver trim, black plastic wheels	£3	£2	£1	50p
5b Golden orange, silver plastic interior, silver trim, headlights and hubs. Made 1970	£5	£4	£2	£1
5c Metallic gold, white interior, gold trim, black superfast wheels. Made 1971	£5	£3	£2	£1
5d Bronzed body, matching trim, white interior, superfast wheels. Made 1972	£3	£2	£1	50p

THE GOLDEN YEARS OF LESNEY

	Mint boxed	Mint unbxd	Good cond	Fair cond
5e Flaw: Golden yellow colour on one side only, no paint on interior	£100	£50		
No 33 Sixth issue Datsun 126X Bright yellow body, bright red base, gold interior, bronze tinted windows, opening boot, red grill and bumpers, length 76mm. Made 1973	£2	£1	50p	25p
6a Golden body, off white interior, bronze or yellow tinted windows, scarlet base bumpers and grill, opening boot. Made 1974	£1	75p	50p	10p
No 33 Seventh issue Datsun 126X Yellow, decals over top sides of car in red and black leaf style, tinted windows, frosted motor window, opening boot, silver hubs, red base, length 76mm, 64:1. Made 1976				
No 34 First issue Volkswagen panel truck Blue body, black base, 'Matchbox International' decal on sides, silver headlights, bumpers and VW on front. Made 1956	£10	£5	£3	£2
1a Orange, yellow decal, metal wheels, crimpled axle	£10	£5	£3	£2
1b Yellow decal, metal wheels, crimpled axle	£10	£5	£3	£2
1c Orange or orange yellow decals on sides, silver trim, grey plastic wheels, crimpled or round axle	£10	£5	£3	£2
1e Yellow decal, grey plastic wheels, round axle	£7	£4	£2	£1
1d Yellow decal, small grey plastic wheels	£6	£5	£3	£2
No 34 Second issue Volkswagen microvan Light blue, orange decals on sides, silver trim, 'International' blue rear bumper, silver plastic wheels, length 56mm, OO scale, 75:1. Made 1961	£10	£5	£2	£1
2a Dark or medium blue or bright sky blue. Made 1961	£7	£5	£2	£1
2b Red or deep maroon body, details as above, rare model made for International year.	£50	£15	£5	£3
No 34 Third issue Volkswagen camping car Light green body, large flat sunroof, darker green interior, tinted windows, silver headlights and VW on front,				

	Mint boxed	Mint unbxd	Good cond	Fair cond
length 64mm, 61:1, OO scale. First casting of this model has raised rear section of seats, solid seat casting held by two rivets on base. Only a few made, worth more than most models. Made 1962	£100	£50	£10	£5
3a Grey plastic wheels, round axle, thin box inside door	£10	£5	£3	£2
3b Green, lighter green interior, thick box inside door, right hand side, round axle knobby tread. Made 1963	£10	£5	£3	£2
3c Dark green body, black interior, box on left hand side door, black plastic wheels, knobby treads	£50	£10	£5	£3
3d Black or green interior, silver trim, grey or black plastic wheels. Second seat casting with the seat section reversed, raised section in front, open space under seats which are held by two rivets on base. Made 1965	£10	£5	£3	£2
No 34 Fourth issue Volkswagen camping car Pale green, darker interior, light green roof, thin box on right hand side, black plastic wheels, fine treads Made 1960	£6	£5	£3	£1
4a Green, tinted roof, green interior, silver trim outside on bumpers, headlights and VW. Seat casting, front section under seats open, no end sections, both seats and table cast to base, no rivets, snap-in base.	£10	£7	£5	£3
Flaw: Paint missing on doors, flat box inside door	£50	£10	£5	£3
No 34 Fifth issue Volkswagen camper Silver body, red interior, orange tinted windows, short raised sunroof with small window, silver trim, two rivets in front of base, black plastic wheels, length 65mm, 66:1. Made 1967	£10	£7	£5	£3
5a As above with licence plate	£10	£7	£5	£3
5b As above with orange interior, orange tinted windows, short raised sunroof window, licence plate and tow slot	£7	£5	£3	£2

THE GOLDEN YEARS OF LESNEY 35

	Mint boxed	Mint unbxd	Good cond	Fair cond

5c Darker silver grey body, orange interior, short raised sun roof. Made 1969 — £5 £3 £2 £1

5d Light metallic blue, silver trim, lights, bumpers and headlights, superfast or non-superfast wheels. Made 1970 — £50 £10

Flaw: Front wheels superfast, back wheels black plastic non superfast, body almost blue-white with the silver appearing in places, orange interior, very rare. Made 1970 — £250 £50

No 34 Sixth issue Formula 1 racing car Pink, white driver, silver motor, sticker decals on hood, number 16 in black, white and yellow flashes, wide superfast racing wheels. Made 1971 — £2 £1 75p 50p

6a Mauve, silver motor and trim, white driver, silver exhausts, plastic windscreen, wide racing wheels — £2 £1 50p 25p

No 34 Seventh issue Formula 1 racing car Yellow, silver motor, white driver, blue decals with arrowhead on bonnet, number 16, wide superfast racing wheels, round pin axles. Made 1974 — £3 £1 75p 50p

7a Darker yellow or mustard, dark blue arrow on bonnet, silver trim. Made 1975 — £3 £1 75p 50p

No 34 Eighth issue Vantastic Red, white interior, white base, white bumpers and grill, silver motor, blue tinted windscreen, wide superfast wheels, length 74mm, 65:1. Made 1976

Flaw: Paint completely missing from body — £25 £15

No 35 First issue Marshall horse box Red cab and base, tan box, silver grill and headlights, drop-down door on right side of box, metal wheels, TT scale 118-1. Made 1956 — £10 £6 £3 £2

1a Dark or light tan box, metal wheels, crimpled axle — £10 £5 £3 £2

1b Light or dark tan box, small grey plastic wheels, crimpled axle — £7 £5 £3 £2

1c Light or dark tan box, small dark or light grey plastic wheels, round axle — £7 £5 £3 £2

1d Light or dark tan box, small black plastic wheels, knobby treads — £5 £4 £3 £2

1e Light or dark tan box, small black plastic wheels, fine treads. Very rare model — £25 £15

No 35 Second issue horse box truck Pink cab, blue plastic wheels, light or dark chocolate coloured body, silver bumpers and headlights, drop down door at right hand side, length 54mm, 48:1, TT scale. Made 1963 — £15 £10 £7 £5

2a Mustard coloured box, deep maroon cab, grey plastic tyres, knobby or fine treads. Made 1964 — £10 £7 £5 £3

2b Flaw: Horse box with no windows, part of door missing — £25 £10

No 35 Third issue Snowtrac tractor Red body, silver grill, base and tow hook, green tinted windows, Snowtrac decals in white, six small plastic rollers with green rubber treads, length 56mm 164:1. Made 1965 — £20 £15 £10 £5

3a Large or small Snowtrac decals in white letters, black plastic rollers — £20 £15 £10 £5

3b Deep red, Snowtrac lettering cast on sides, black plastic rollers — £10 £7 £5 £3

3c No decals or lettering, green plastic rollers — £25 £20 £10 £5

3d No decals or lettering, black plastic rollers — £10 £7 £5 £3

No 35 Fourth issue Snowtrac tractor Dark red, salmon coloured tracks, blue grill and tow hook, green tinted windows, Snowtrac cast on side, length 56mm, 64:1. Made 1967 — £25 £15 £10 £5

4a Silver plastic rollers, body as above — £25 £20 £15 £10

4b Pink, green tinted windows, silver trimmed roof, silver grill and tow hook, black snow tracks. Made 1969 — £25 £20 £10 £5

No 35 Fifth issue Merryweather Marquis fire engine Metallic red, blue windows and dome lights, grey

THE GOLDEN YEARS OF LESNEY

	Mint boxed	Mint unbxd	Good cond	Fair cond
plastic base, white ladder on top, sticker decals on sides, London Fire Service in white, blue and red, length 76mm, 86:1, superfast wheels. Made 1970	£10	£7	£5	£3
5a Metallic red letter A inside circle on front base	£10	£7	£5	£3
5b Bright red, letter A inside circle, metallic red letter A on flat base	£7	£5	£3	£2
5c Bright red, letter A inside circle, 'PAT APP' now behind rear axle	£10	£7	£5	£3
No 35 Sixth issue Fandango White, orange interior, flowered decal on roof behind, orange pointed arrowhead flash on front bonnet, number 35 in dark blue outline, length 76mm, 60:1. Made 1975	£1	75p	50p	25p
6a Off white, red interior, white flowered decal on rear, dull orange pointed arrow, number 35 with light blue outline, length 75mm. Made 1976	£1	75p	50p	25p
No 36 First issue Austin A50 Light turquoise, black base, silver grill, headlights and bumpers, red tail lights, metal wheels, crimpled axle, OO scale. Made 1956	£10	£7	£5	£3
1a Blue green, metal wheels, crimpled axle	£10	£7	£5	£3
1b Light turquoise or blue green, sun visor, metal wheels, crimpled axle	£5	£4	£3	£2
1c Blue green or light blue green body, sun visor, grey plastic wheels, crimpled axle	£5	£3	£2	£1
1d Light blue, sun visor, grey plastic wheels, crimpled axle	£4	£2	£1	75p
No 36 Second issue Austin A50 Red, silver grill, bumper bars, headlights, tow hook and silver lined trim on body, metal or grey plastic wheels. Very rare model	£500	£100		
No 36 Third issue motor scooter and sidecar Light metallic green, three black plastic wheels, knobby treads, length 48 mm, 41:1, O scale?	£25	£15	£10	£5
3a Dark metallic green, black plastic wheels, knobby treads	£25	£15	£10	£5

	Mint boxed	Mint unbxd	Good cond	Fair cond
3b Light or dark metallic green, black plastic wheels, fine treads	£25	£15	£10	£5
No 36 Fourth issue Lambretta scooter and sidecar Light blue cycle and sidecar, grey plastic wheels, fine and knobby treads, length 49 mm, 41:1, O scale. Made 1963	£25	£15	£10	£5
4a Dark blue, otherwise details as above. Made 1964	£25	£15	£10	£5
4b Very light metallic green. Made 1965	£20	£10	£7	£5
Flaw: handlebars completely missing	£50	£25		
No 36 Fifth issue Opel Diplomat Light apple green, white interior, clear windows, opening bonnet, silver motor, tow hook, black plastic wheels. Made 1966	£7	£5	£3	£2
5a Metallic gold. Made 1967	£7	£5	£3	£2
5b Bright yellow body. Made 1968	£7	£5	£3	£2
5c Silver or grey motor with licence plate and tow slot	£7	£5	£3	£2
5d Dark metallic gold, silver motor, superfast wheels. Made 1970	£5	£3	£2	£1
5e Light, medium and dark metallic gold bodies, long or short steering wheels, superfast wheels	£5	£4	£3	£2
No 36 Sixth issue Hot Rod Draguar Metallic red, white plastic interior, clear plastic dome, silver exhaust and motor, silver grill at rear, silver motor and wide superfast wheels, length 70mm, 63:1. Made 1971	£2	£1	75p	50p
6a Lighter red, pink motor and exhausts	£2	£1	75p	50p
6b Pink body, cream interior	£1	75p	50p	25p
6c Turquoise body, white interior, silver motor	£2	£1	75p	50p
No 36 Seventh issue formula 5000 Red, dark blue and white decals, silver base, black number 3 on bonnet and formula 5000 on tail and nose in black and white decals, white line trim and blue flashes, wide superfast wheels, length 74mm, 58:1. Made in 1976				

THE GOLDEN YEARS OF LESNEY 37

	Mint boxed	Mint unbxd	Good cond	Fair cond

No 37 First issue Coca Cola lorry Orange yellow, large lettered decal at top of load and rear, silver grill, headlights and front bumper, open base with metal rear fenders, metal wheels, crimpled axle, HO scale, 89:1. Made 1956 — £100 £50 £15 £10

1a Yellow, orange yellow, small letter decals at top of load and on body — £100 £50 £15 £10

No 37 Second issue Coca Cola lorry Orange yellow and yellow, Coca Cola decals at top of load and at rear, even case load, silver grill, headlights and front bumper, open base with metal rear fenders, metal wheels, crimpled axle. Made 1957 — £10 £7 £5 £3

2a Details as above with grey plastic wheels — £25 £15 £10 £5

No 37 Third issue Coca Cola lorry Orange yellow and yellow, black base, black rear fenders, even case load, silver grill, headlights and front bumper, grey plastic wheels, crimpled axle, Coca Cola decals on top and at rear, length 61 mm, 89:1, HO scale. Made 1965 — £7 £5 £3 £2

3a Orange yellow, grey plastic wheels, round axle — £7 £5 £3 £2

3b Yellow with grey plastic wheels or round axle with light grey plastic wheels — £7 £5 £3 £2

3c Yellow body, rear decals cut short or long from top to bottom, black plastic wheels, knobby treads — £7 £5 £3 £2

3d Silver plastic wheels; a very rare model made for the International Show, a very special year for Lesney. Last issue made in 1966. — £50 £25 £10 £5

No 37 Fourth issue cattle truck Orange body and chassis, green tinted windows, grey plastic box with white plastic cattle, small wing supports under rear overhanging under tail gate, silver bumpers, grill and hubs, length 65mm, 86:1. Made 1967 — £10 £7 £5 £3

4a Yellow body and chassis, dark grey cattle box. Made 1968 — £10 £7 £5 £3

4b Dark brown body, very dark cattle box, no wing supports, silver plastic base and grill, — £10 £7 £5 £3

4c Light or dark grey box, metal base and grill, with or without tow slot — £7 £5 £3 £2

4d Dark orange body, dark green box, superfast wheels. Made 1970 — £7 £5 £3 £2

4e Orange mustard body, black box with animals, superfast wheels; rounded or half rounded middle bar ends on rear box. Very rare model — £15 £7 £5 £3

No 37 Fifth issue Soopa Coopa Bright blue, yellow interior, yellow tinted plastic cover, silver grill and rear with white base, side windows at rear, wide superfast racing wheels, length 70 mm, 60:1. Made 1973 — £2 £1 50p 25p

5a Dark blue body, orange interior. Made 1974 — £2 £1 50p 25p

5b Royal blue body. Made 1975 — £2 £1 50p 25p

5c Bright mauve metallic body, yellow interior, red flowered decal on roof, length 74mm. Made 1976

37, Coca Cola lorry. 38, Refuse collector truck

38 THE GOLDEN YEARS OF LESNEY

	Mint boxed	Mint unbxd	Good cond	Fair cond

No 38 First issue Karrier refuse collector Dark grey body, metal wheels, decal with yellow words, silver grill, headlights and front bumper, cleansing dept sign on sides, decal with yellow words, opening between cab and body, HO scale, 87:1. Made 1956 — £10 £7 £5 £3

1a Grey cab and body, orange decal, metal wheels, crimpled axle. Made 1957 — £10 £7 £5 £3

1b Grey cab and body, filled in to a point at the top, metal wheels, crimpled axle, orange decal — £7 £5 £3 £2

1c Yellow decal, grey plastic wheels, crimpled axle — £7 £5 £3 £2

1d Grey body, orange decal, grey plastic wheels, round axle — £7 £5 £3 £2

1e Light grey body, light grey plastic wheels, round axle, black decal, rare model — £25 £15 £10 £5

1f Very light silver cab and body, yellow decal, black outline, grey plastic wheels, round axles — £25 £15 £10 £5

No 38 Second issue refuse truck silver blue body and cab, silver headlights, grill and bumper, cleansing dept decals in silver with yellow tint, wider opening than other models between cab and body, black base, silver plastic wheels, length 64mm. Last model made 1962 — £50 £25 £10 £5

Flaw: no paint on cab and slight silver with blue tint on body, missing decals, no silver on headlights or bumper — £250 £100

No 38 Third issue Vauxhall Victor Estate Royal blue, black base and tow hook, opening rear doors, silver grill, headlights and bumper, grey plastic wheels, knobby treads, clear plastic windows. Made 1963 — £25 £20 £10 £5

3a Yellow body and interior, grey plastic wheels, other details as above — £10 £5 £3 £2

3b Green interior, small grey plastic wheels, knobby treads — £7 £5 £3 £2

3c Green interior, large grey plastic wheels, fine treads — £7 £5 £3 £2

No 38 Fourth issue Vauxhall Victor Estate Light yellow, green interior, silver bumper, headlights and grill, opening rear door, black plastic wheels with fine treads, length 64mm, 71:1, OO scale. Made 1966 — £7 £5 £3 £2

4a Yellow body, red interior, silver plastic wheels, knobby treads — £7 £5 £3 £2

4b Red interior, small or large black plastic wheels, fine treads — £5 £3 £2 £1

4c White interior, small or large black plastic wheels, fine treads — £25 £15 £10 £5

4d Red or dark maroon body, red interior, silver plastic wheels, knobby treads. Last model made 1966, but models in normal colours in 1967 — £250 £100

No 38 Fifth issue Honda motor cycle with trailer Metallic blue green cycle, orange trailer, wire wheels, Honda decal on trailer, black plastic wheels, length 76mm, 41:1. Made 1968 — £10 £7 £5 £3

5a Above model with sticker decals — £7 £5 £3 £2

5b Red orange trailer, no decals, short axle channel — £10 £7 £5 £3

5c Yellow trailer, large lettered decal, full axle channel, blue motor cycle, brace on inside channel — £7 £5 £3 £2

39, Ford Zodiac convertible. 40, Bedford tipper truck

	Mint boxed	Mint unbxd	Good cond	Fair cond

5d Yellow trailer, blue motor cycle, large lettered decal with full axle channel, braces on both sides. Made 1969 — £7 £5 £3 £2

5e Blue motor cycle, yellow trailer, sticker decals, superfast wheels. Made 1970 — £5 £3 £2 £1

5f Green motor cycle, yellow trailer, superfast wheels. Made 1971 — £5 £3 £2 £1

5g Light pink motor cycle, golden spoked wheels, yellow trailer, superfast gold trim wheels on trailer, very rare. Made 1972 — £50 £25 £15 £10

No 38 Sixth issue Stingeroo Purple frame and seat, silver forks and silver seat support, silver engine, wide superfast wheels, length 82mm, 50:1. Made 1973 — £1.50 £1 75p 50p

No 38 Seventh issue Stingeroo Lilac, sky blue forks and handlebars, silver hubs and motor, plastic head behind seat and silver exhaust, length 78mm. wide, superfast wheels at rear and smaller wheels at front. Made 1974 — 60p 50p 20p 10p

7a Very deep lilac, brighter orange between forks, otherwise details as above. Made 1975 — 60p 50p 20p 10p

7b Lighter body, darker handlebars, white head silver trim. Made 1976

No 39 First issue Ford Zodiac convertible Bronze, tan driver, light green interior, green tow hook, green base, silver grill, headlights and bumpers, red tail lights, metal wheels, crimpled axle, OO scale, 71:1. Made 1956. — £10 £7 £5 £3

1a Light peach body, green interior and base, metal wheels, crimpled axle — £7 £5 £3 £2

1b Light pink body, turquoise interior and base, grey plastic wheels, crimpled or round axle — £7 £5 £3 £2

1c Dark pink, turquoise interior and base, grey plastic wheels, round axle — £5 £4 £3 £2

1d Light pink, turquoise interior and base, silver plastic wheels, round axle — £7 £5 £3 £2

THE GOLDEN YEARS OF LESNEY 39

	Mint boxed	Mint unbxd	Good cond	Fair cond

No 39 Second issue Ford convertible Bright rust with silver trim along body, blue interior and base, silver bumper and grill and headlights, yellow driver, length 68mm, 71:1, OO scale, red tail lights, silver tow hook — £25 £15 £10 £5

No 39 Third issue Pontiac convertible Purple body, light and dark, red steering wheel, white interior, silver plastic wheels, grill and headlights and silver bumpers at front and rear, green tinted windscreen, length 70 mm, 80:1, OO scale. Made 1962 — £15 £10 £5 £3

3a Purple, cream interior; red steering wheel and base; silver plastic wheels, knobby treads, otherwise as above — £7 £5 £3 £2

No 39 Fourth issue Pontiac convertible Golden yellow, cream interior, red steering wheel and base, silver grill, headlights and bumpers, tow hook and tinted windshield, silver plastic wheels, knobby treads, length 70mm, 80:1, OO scale. Made 1964 — £7 £5 £3 £2

4a Golden yellow, cream or off-white interior, red steering wheel and base, silver plastic wheels, fine treads. Made 1965 — £7 £5 £3 £2

4b Yellow, off white interior, black base, silver plastic wheels, knobby treads. Made 1966 — £5 £4 £3 £2

4c Yellow, off white interior, black base, thick or thin grey plastic wheels, knobby treads — £5 £4 £2 £1

4d Yellow, off white interior, black base and plastic wheels, fine treads — £5 £4 £2 £1

4e Yellow, off white interior, middle front grill silver only, black plastic wheels — £10 £7 £5 £3

4f Very light yellow, off white or cream interior, front grill silver only. Last model made 1967 — £10 £7 £5 £3

No 39 Fifth issue Ford tractor Blue, black steering wheel, orange hubs, black plastic tyres; length 54 mm, 56:1. Made 1968 — £10 £7 £5 £3

5a Light blue, short stack, yellow hubs — £7 £5 £3 £2

5b Dark blue, yellow orange hood and front grill, short or tall stacks, black plastic tyres — £7 £5 £3 £2

THE GOLDEN YEARS OF LESNEY

	Mint boxed	Mint unbxd	Good cond	Fair cond

No 39 Sixth issue Ford tractor Deep purple, bright orange hood and grill, purple axles, orange hubs, black plastic tyres. Made 1969 — £7 £5 £3 £2

6a Deep purple, yellow hood and grill, short or long stacks — £7 £5 £3 £2

6b Very light blue, yellow hubs, hood and grill, silver headlights. Made 1970 — £6 £5 £3 £2

Flaw: missing drivers seat and steering wheel, no stack — £25 £15 £10 £5

No 39 Seventh issue Clipper Purple, yellow interior, clear plastic window, silver trim with green inset at front, large wide superfast wheels at rear and small superfast at front, length 83 mm, 59:1 — £1 75p 50p 25p

7a Metallic light pink, orange interior, tinted window, blue inset at nose, silver trim, superfast wheels. Made 1974 — £1 75p 50p 25p

7b Very deep metallic purple, 77 mm. Made 1976

No 40 First issue Bedford tipper truck 7 ton type Red cab and chassis, light tan dumper, silver grill, headlights and bumper, tail gate swings open, metal wheels, crimpled axle, length 93:1, HO scale, Made 1956 — £15 £10 £7 £5

1a Red cab and chassis, dark tan dumper, metal wheels, crimpled axle — £10 £7 £5 £3

1b Red cab, darker red chassis light tan dumper, grey plastic wheels, crimpled axle — £7 £5 £3 £2

1c Dark red cab, lighter chassis, light or dark tan dumper, grey plastic wheels, round or crimpled axle — £7 £5 £3 £2

1d Dark red cab, very dark chassis, dark tan dumper, small grey plastic wheels in front and larger grey plastic wheels with knobby treads at rear. Very rare. Made in 1960 — £100 £50

No 40 Second issue Royal Tiger long distance coach Silver grey, silver grill and bumpers, red tail lights, four plastic wheels, length 75 mm, 145:1. Made 1961 — £25 £15 £10 £5

	Mint boxed	Mint unbxd	Good cond	Fair cond

2a Red tail lights, silver plastic wheels, knobby treads — £10 £7 £5 £3

2b Silver blue, blue front grill, red tail lights, silver plastic wheels, knobby treads. Made 1962 — £7 £5 £3 £2

2c Dark silver blue, red tail lights, black plastic wheels, fine treads, length 76 mm. Made 1963 — £7 £5 £3 £2

2d Dark or light body, black plastic wheels, rivet on gear base holding bumper. Made 1965 — £7 £5 £3 £2

No 40 Third issue Royal Tiger long distance coach Special model made for 1966 International: 3-tone silver grey, silver light blue and silver dark blue, dark tinted windows, red base and tail lights, special rear fins and prominent nose, silver plastic wheels, length 76 mm, 45:1, very streamlined, only very few in existence — £250 £100

3a Light or dark silver blue, blue tinted windows blue hubs and red tail light, light bluish grill and bumpers, black plastic tyres. Made 1967 — £10 £7 £5 £3

No 40 Fourth issue hay trailer Deep blue, orange plastic racks and hubs, black plastic tyres, length 86 mm, 56:1. Made 1968 — £7 £5 £3 £2

4a Blue, yellow plastic racks and hubs, rack pins go into base with six sided brace under rear axle — £7 £5 £3 £2

4b Rack pins go into base; round brace under rear axle — £5 £4 £3 £2

4c Rack pins go into blocks under base with round brace under axle — £5 £3 £2 £1

4d Racks fit only in slots. Made 1970/71 — £5 £4 £2 £1

No 40 Fifth issue Guildsman Shocking pink, white interior, green tinted windows, metal base, sticker decals on hood, length 76 mm, 64:1 — £2 £1 75p 50p

5a Dark pink body, off white interior — £1 75p 50p 25p

5b Light or dark metallic pink. Made 1975 — £1 75p 50p 25p

No 40 Sixth issue Guildsman Metallic bronze, yellow decals on front, centre and rear, bordered with orange line or dark red line with number

THE GOLDEN YEARS OF LESNEY

	Mint boxed	Mint unbxd	Good cond	Fair cond
40 on bonnet in blue, silver trim, wide superfast wheels. Made 1976				
No 41 First issue D-type Jaguar Dark green, tan driver, black base with open scoop. metal wheels crimpled axle	£10	£7	£5	£3
1a Number 41 decal on front and rear, metal wheels, crimpled axle	£10	£7	£5	£3
1b Decals on front and rear, grey plastic wheels, crimpled axle	£5	£4	£3	£2
No 41 Second issue D-type Jaguar Dark green, tan driver, black base, number 41 decals on front and rear, grey plastic wheels. Made 1960	£10	£7	£5	£3
2a Light green, yellow driver, open scoop, grey plastic wheels, crimpled or round axles. Made 1961/62	£6	£5	£3	£2
2b Closed scoop, large grey plastic wheels, round axle. Made 1963	£5	£4	£2	£1

	Mint boxed	Mint unbxd	Good cond	Fair cond
2c Closed scoop, large or small light grey plastic wheels, round axle	£5	£4	£2	£1
2d Closed scoop, large grey plastic wheels, fine treads	£5	£4	£2	£1
2e Number 41 and 5 decal, closed scoop, wire wheels with black plastic tyres	£7	£5	£3	£2
2f Black, tan driver, red hubs, closed scoop, length 64 mm, green almost completely disappeared. Made 1965	£100	£50	£15	£5
No 41 Third issue Ford G T Clear white body and interior, silver rear motor, streamlined decals on bonnet, yellow plastic hubs, black plastic tyres, length 67 mm, 67:1. Made 1966	£7	£5	£3	£2
3a Yellow body and hubs, dark blue number 6 decal	£4	£3	£2	£1
3b White, yellow hubs, dark blue number 6 decal, also comes with tow slot	£4	£3	£2	£1
3c White, yellow hubs, dark blue number 6 decal, tow slot and licence plate	£4	£3	£2	£1
No 41 Fourth issue Ford G T Silver grey, grey number 6 decal, silver plastic hubs, black plastic wheels, knobby treads, black base, length 72 mm, 61:1. Made 1969	£7	£5	£3	£2
4a White, yellow hubs, dark, medium and very light blue sticker decals	£4	£3	£2	£1
4b White, yellow hubs, light blue number 9 sticker decals	£15	£10	£5	£3
4c White, dark, medium and light blue number 6 sticker decal, superfast wheels	£5	£4	£3	£2
4d White, green base, medium blue number 6 sticker decal, unpainted motor, superfast wheels	£15	£10	£5	£3
4e White, green base, medium blue number 6 sticker decals, blue or dark blue tinted motor	£4	£3	£2	£1
4f White, green base, silver motor, light blue number 6 sticker decal, superfast wheels	£5	£4	£3	£2
4g White, green base, silver motor, light and dark blue number 9 sticker decal, superfast wheels	£25	£15	£10	£5

41, D Type Jaguar. 42, Bedford Evening News van

42 THE GOLDEN YEARS OF LESNEY

	Mint boxed	Mint unbxd	Good cond	Fair cond
4h White, very bright green base, light blue number 6 sticker decal, dark blue tinted motor, superfast wheels	£5	£4	£3	£2
No 41 Fifth issue Ford G T Metallic red, black base, tinted windows, silver motor, medium blue number 6 sticker decal, wide superfast wheels. Made 1972	£7	£5	£3	£2
5a Metallic red, yellow or light and dark green base, silver motor, medium blue number 6 sticker decal	£5	£4	£2	£1
5b Metallic red, green base, silver motor, medium or dark blue number 6 decal, wide racing tyres. Made 1972	£4	£3	£2	£1
Flaw: Ford G T with no seats, red number 9 decal, metallic red body	£25	£15		
No 41 Sixth issue Siva Spyder Pink, off-white interior, white base, wide superfast wheels, length 84 mm, 54:1. Made 1973	£2	£1	75p	50p
6a Lighter body, black section in centre of vehicle	£1	75p	50p	25p
6b Very light pink. Made 1975	£1	75p	50p	25p
No 41 Seventh issue Siva Spyder Royal blue, red, white and blue decals on back, top and bonnet, stars and stripes with two red, white and blue circles on roof, wide superfast wheels, length 77 mm, 54:1. Made 1976				
No 42 First issue Bedford Evening News van Yellow, black base, silver grill, headlights and bumper, top decal with red background and white letters 'First with the News', Evening News body decal in red, door decal in white, black and red 'Football results, Evening News', metal wheels crimpled axle, length 70:1, OO scale. Made 1956	£50	£25	£7	£5
1a decals on one of first issues had 'Football results and Evening News' upside down, Very bad flaw	£1000	£250		
1b Grey plastic wheels, crimpled axle	£10	£5	£3	£2

	Mint boxed	Mint unbxd	Good cond	Fair cond
1c Grey plastic wheels, round axle	£10	£5	£3	£2
1d Black plastic wheels, knobby or fine treads	£7	£5	£3	£2
Flaw: black plastic wheels, knobby treads on front axle, fine treads on rear, Evening News decals missing on sides. Made 1960	£50	£25		
No 42 Second issue newspaper delivery truck Dark brown, silver base, headlights and bumper bars, white decal 'first with the news' on top, bright orange 'Evening News' decal on side, black and purple 'Football Results and Evening News' decal on door, grey plastic wheels, length 56 mm, 70:1, OO scale. Made 1961	£25	£10	£5	£3
2a Light yellow body, otherwise details as above, length 57 mm	£7	£5	£3	£2
2b Bright orange, no silver trim, knobby treads, length 57 mm. Made 1963	£25	£5	£3	£2
2c Light yellow, silver trim, bright red Evening News sign on side, all black lettering decals on door. Made 1964	£25	£10	£5	£3
No 42 Third issue Studebaker station wagon Light blue body, white interior, grey plastic man and dogs, sliding sunroof panel, silver trim, black plastic wheels, length 76 mm, 66:1. Made 1965	£10	£5	£3	£2
3a Dark blue body, blue sliding panel	£5	£4	£3	£2
3b Dark or light blue, light blue sliding panel and tow slot, silver trim	£5	£3	£2	£1
3c Light blue, silver grey interior, dark blue trim along sides and bottom and roof panels, special curved sliding sun roof, white plastic man and dogs, silver bumpers, headlights and grill, length 76 mm. Exquisite model made in 1966 for special International	£25	£10	£5	£3
3d Very dark, plain body, no trim. Made 1967	£5	£4	£3	£2
3e Plain sky blue body, blue hubs, bumpers headlights and grill, normal sliding sun roof. Made 1968	£5	£4	£3	£2
3f Green, white interior, silver trim. Made 1969	£10	£5	£3	£2

	Mint boxed	Mint unbxd	Good cond	Fair cond

No 42 Fourth issue iron fairy crane Orange body and crane, yellow hook, orange hydraulic sleeve, yellow orange base, black plastic superfast wheels, length 76 mm, 80:1. Made 1970 £5 £4 £3 £2

4a Red body, black plastic wheels £4 £3 £2 £1

4b Red or orange, silver superfast wheels £5 £3 £2 £1

4c Red orange, orange yellow base, lime green crane, superfast wheels £5 £4 £3 £2

No 42 Fifth issue tyre fryer Bright blue, yellow seats, silver trim, silver mounted engine with dark brown top mounted at rear, large wide superfast wheels at back, small superfast wheels at front, length 76 mm, 52:1. Made 1973 £2 £1 75p 25p

5a Dark blue, dark cover, orange seat, silver mounted engine, large wheels at rear and small at front. Made 1974 £1 75p 50p 25p

5b Medium blue, yellow seat, silver trim, length 76 mm, 52:1. Made 1976

No 43 First issue Hillman Minx Apple green, black base, silver grill, headlights and bumpers, red tail lights, open windows, tow hook, metal wheels, 64:1. Made 1957 £10 £5 £3 £2

1a Dark green, silver trim, metal wheels, crimpled axle £25 £10 £5 £2

1b Dull blueish grey lower body, light grey top, metal wheels, crimpled axle. Made 1958 £10 £5 £3 £2

1c Dull blue grey, light grey top and back, grey plastic wheels, crimpled axle. Made 1959 £7 £5 £2 £1

1d Light blue, cream top and back, tow hook, grey plastic wheels, crimpled axle, knobby treads £10 £5 £2 £1

1e Light blue, lighter blue on roof, grey plastic wheels, crimpled axle £50 £15

1f Light green, cream roof, length 72 mm, 64:1, OO scale,. Made 1962 £7 £5 £2 £1

1g Turquoise, light cream top and back, grey plastic wheels, round axle. Made 1962 £7 £5 £2 £1

43, Hillman Minx. 44, Rolls Royce Silver Cloud

	Mint boxed	Mint unbxd	Good cond	Fair cond

No 43 Second issue Aveling Barford tractor shovel Fawn body, fawn driver, moveable scoop, four large black plastic wheels, length 67 mm, 97:1, TT scale. Made 1963 £10 £7 £5 £2

2a All yellow body and driver £10 £7 £5 £2

2b Dark yellow, red driver, four large black plastic wheels. Made 1966 £10 £7 £5 £2

2c Mustard, red driver, red scoop. Rare model made in 1966/67 £25 £10 £5 £3

2d All yellow, red driver, blue hubs on large blue coat plastic tyres, knobby treads. Made 1967 £10 £5 £2. £1

No 43 Third issue Aveling Barford tractor shovel Half scoop and missing driver with only three tyres with knobby treads and one with smooth treads. Bad factory flaw £100 £25

No 43 Fourth issue pony trailer Yellow, green base and tow bar, clear windows, 2 horses with grey plastic tail gate, length 67 mm, 63:1. Made 1968 £6 £4 £2 £1

4a Tan base, tow bar, grey tail gate £5 £3 £2 £1

4b Green base, tow bar, off white grey tail gate, two grey horses £7 £5 £2 £1

THE GOLDEN YEARS OF LESNEY

	Mint boxed	Mint unbxd	Good cond	Fair cond
4c Green base, tow bar, light grey tail gate, superfast wheels	£4	£3	£2	£1
4d Light green base, superfast wheels. Made 1970	£4	£3	£2	£1
4e Golden yellow trailer, green tow bar, dark brown tail gate, two white horses. Made 1971/72	£4	£3	£2	£1
Flaw: paint completely missing from trailer	£100	£50		
No 43 Fifth issue Dragon Wheels Very light green body lifts up to reveal silver plastic motor and interior, red seats, dragon wheel decals on sides, wide plastic superfast wheels, length 70mm. 59:1. Made 1973	£2	£1	50p	25p
5a Dark green, otherwise details as above	£2	£1	50p	25p
5b Dark green, slightly altered silver motor, badly warped, very cheap-looking appearance. Made 1975/76				
No 44 First issue Rolls Royce Silver Cloud Metallic blue, silver grill, headlights and bumpers, black base, red tail lights, metal wheels, crimpled axle, OO scale, 80:1. Made 1957	£10	£7	£5	£3
1a Metallic silver blue, grey plastic wheels, crimpled axle	£7	£5	£2	£1
1b Metallic blue or metallic dark or silver blue, grey plastic wheels, round axle	£5	£4	£2	£1
1c Metallic dark blue, silver plastic wheels, knobby treads	£5	£4	£2	£1
1d Metallic silver blue, metallic dark or metallic medium blue, blue tail lights, silver plastic wheels	£5	£4	£2	£1
1e Metallic blue, red or blue tail lights, silver plastic wheels, finer knobby treads	£5	£4	£2	£1
No 44 Second issue Rolls Royce Phantom VR Metallic silver grey body, silver grill, bumper bars and opening boot, golden yellow interior, clear windows, black plastic wheels, length 72 mm, 74:1, OO scale Made 1964	£7	£5	£2	£1
2a All silver body, off white interior, black plastic wheels	£25	£15	£10	£5
2b Metallic pinkish tan, golden orange interior, blue grill and bumper bars at front and rear, grey plastic wheels, knobby treads. Made 1966	£15	£10	£5	£3
2c Very light metallic tan, black plastic wheels	£5	£4	£2	£1
2d Deep purple metallic body, off white interior, silver grill, bumper bars and opening boot, blue-black plastic wheels. Made 1967	£25	£10	£5	£3
Flaw: Paint missing from one side, interior missing	£100	£25		
No 44 Third issue refrigerator truck Red cab and chassis, green box, green tinted windows on cab, grey plastic grill, headlights and quarter base, black plastic wheels, length 76 mm, 86:1	£10	£7	£5	£3
3a Red cab, turquoise box, red chassis, very light rear door. Made 1969	£5	£4	£2	£1
3b Red cab, turquoise box, dark rear door, tow slot, black plastic wheels. Made 1969	£5	£4	£2	£1
3c Red cab, blue green box, superfast wheels. Made 1970	£5	£4	£2	£1
No 44 Fourth issue refrigerator truck Yellow cab with chassis, deep red box, superfast wheels. Made 1971	£10	£5	£3	£2
4a Light yellow cab and chassis, deep or light red box, superfast wheels. Made 1972	£7	£5	£2	£1
Flaw: truck without door, no paint on box	£50	£25		
No 44 Fifth issue Boss Mustang Yellow, off-white interior, yellow tinted windows, black opening bonnet, white plastic engine, silver grill and bumpers at front and rear, silver base, superfast wheels, length 76 mm, 86:1. Made 1973	£2	£1	50p	25p
No 44 Sixth issue Boss Mustang Dark orange body, orange tinted windows, silver bumper, grill and base, dark blue opening bonnet with silver emblem, superfast wheels, length 76 mm, 86:1. Made 1974	£1	75p	25p	15p
6a Deep yellow, black bonnet and silver emblem, grey superfast wheels, length 75mm. Made 1975	£1	50p	25p	15p

THE GOLDEN YEARS OF LESNEY

	Mint boxed	Mint unbxd	Good cond	Fair cond

6b Deep mustard, silver trim, black opening bonnet, silver base, orange tinted windows. Made 1976

No 45 First issue Vauxhall Victor saloon Yellow body, silver grill, headlights and bumpers, red tail lights, green tinted windows, metal wheels, metal tow hook, OO scale, 72:1. Made 1957 — £10 £5 £2 £1

1a Deep golden yellow, silver trim on sides, metal wheels, crimpled axle, no windows — £7 £5 £2 £1

1b No windows, side trim missing, metal wheels, crimpled axle — £25 £10

1c No windows, grey plastic wheels, crimpled axle — £5 £4 £2 £1

1d With or without windows, grey plastic wheels, round axle — £5 £4 £2 £1

1e Windows, yellow rear bumper, grey plastic wheels, round axle — £10 £4 £2 £1

1f Windows, yellow rear bumper, large grey plastic wheels, round axle — £7 £4 £2 £1

1g Windows, yellow rear bumper, large or very small silver or plastic wheels. Made 1963 — £7 £4 £2 £1

1h Deep or light yellow, small black plastic wheels — £5 £3 £2 £1

1j Large black plastic wheels, very rare — £50 £15

No 45 Second issue Ford Corsair with boat Dark yellow, red interior, tow hook, silver grill, headlights, green plastic roof rack and green boat, silver metal base, black plastic wheels, length 67 mm, 71:1, OO scale — £10 £5 £2 £1

No 45 Third issue Ford Corsair with boat Cream body, blue interior, green roof rack, light green boat, silver blue bumpers, grill, and headlights, silver base, length 67 mm. Made 1966 — £7 £5 £2 £1

3a Silver metal or grey metal base, black plastic wheels. Made 1967 — £5 £3 £2 £1

3b Yellow body, dark green boat, pink interior, silver grill and bumpers. Made in 1969 — £4 £3 £2 £1

No 45 Fourth issue Ford group six Metallic green body, off-white interior, silver rear motor, clear windows, wide racing superfast wheels, number 7 decal on hood, length 76 mm, 56:1. Made 1970 — £5 £4 £2 £1

4a Dark metallic green, silver metal base, round circle number 7 sticker decal — £4 £3 £2 £1

4b Medium metallic green or light metallic green, silver metal base — £4 £2 £1 50p

4c Dark metallic green, black base, round circle and number 9 decal — £50 £25

4d Dark metallic green, black base, round circle decal with side decals in red, white and blue, 'Burmah'. Special issue, very rare model — £25 £15 £5 £3

4e Dark metallic green, black base, square number 7 decal — £10 £5 £2 £1

4f Medium metallic green, black or yellow base, square number 45 decal — £15 £10 £2 £1

45, Vauxhall Victor saloon. 46, Morris Minor

THE GOLDEN YEARS OF LESNEY

	Mint boxed	Mint unbxd	Good cond	Fair cond
4g Metallic green, light yellow or pink base, number 45 decal	£50	£25		
No 45 Fifth issue Ford group six Light green, off-white interior, number 45 decal. Made 1973	£2	£1	75p	50p
5a Metallic pink, number 45 decal. Made 1974	£2	£1	75p	50p
5b Deep metallic red, number 45 decal, last model in series. Made 1975.	£1	75p	50p	25p
No 45 Sixth issue BMW 30 CSL Metallic red, yellow or orange interior, two star decal with black, orange and white ring and orange flash on bonnet, silver grill, bumpers, headlights and hubs, opening doors, length 74 mm, 62:1. Made 1976.				
No 46 First issue Morris Minor Dark blue green, black base, silver grill, headlights and bumpers, red tail lights, metal wheels, crimpled axle, length 72:1, OO scale. Made 1957.	£10	£5	£2	£1
1a Royal blue, metal wheels, crimpled axle. Only a few known in existence	£100	£50	£10	£5
1b Dark green, metal wheels, crimpled axle	£7	£5	£5	£1
1c Dark green, grey plastic wheels, crimpled axle	£7	£5	£2	£1
1d Light, medium and dark blue, grey plastic wheels, round or crimpled axles. Made 1960	£5	£4	£2	£1
No 46 Second issue long distance box van Dark blue, silver grill and headlights and bumper, cream plastic rear sliding door, white lettered decal on sides 'Pickfords Removers and Storers' 'Branches in all large towns', grey plastic wheels and round axle, length 66 mm, TT scale, 111:1. Made 1961	£15	£7	£5	£3
Flaw: no decal on one side	£50	£25		
2a Dark blue, two line decal only, bottom line missing	£25	£10		
2b Dark blue and light blue body, grey plastic wheels, three line decal	£25	£10	£5	£3

	Mint boxed	Mint unbxd	Good cond	Fair cond
2c Dark blue, large silver plastic wheels, three line decal	£10	£5	£2	£1
2d Dark blue, large silver plastic wheels, three line decal	£10	£5	£2	£1
2e Dull green, three line decal, length 67 mm. Made 1963/64	£7	£5	£3	£1
2f Light green, silver plastic wheels. Made 1965	£7	£5	£3	£1
2g Bright blue-green body, black plastic wheels. Made 1966	£10	£5	£3	£1
2h Sea green body. Made 1967	£7	£5	£2	£1
No 46 Third issue Mercedes 300 SE Green, opening doors and boot, metal grill, bumpers and base, tow slot, grey plastic wheels, knobby treads, length 73 mm, 66:1. Made 1968.	£10	£5	£2	£1
3a Green, black plastic wheels	£5	£4	£2	£1
3b Metallic light blue, off white interior, silver grill, headlights and bumper bars. Made 1969	£5	£4	£2	£1
3c Metallic dark or medium blue, black plastic wheels. Made 1969	£5	£4	£2	£1
3d Metallic apricot, silver trim and hubs, superfast wheels. Made 1970	£7	£5	£2	£1
3e Orange metallic body. Made 1971	£7	£5	£2	£1
3f Metallic gold body. Made 1972	£7	£5	£2	£1
Flaw: paint completely missing from body	£75	£25		
No 46 Fourth issue Stretcha Fetcha White body, orange base, off white interior, blue lights on roof of cab, opening rear door, red cross flashes on sides, curved nose, superfast wheels, length 70 mm, 66:1. Made 1973	£2	£1	40p	25p
4a Off white, blue tinted windows. Made 1974/75	£1	50p	25p	10p
4b Bright orange base, dark blue tinted windows. Made 1976				
No 47 First issue Brooke Bond Trojan van Red, black base, silver headlights and grill, red grill,				

THE GOLDEN YEARS OF LESNEY 47

	Mint boxed	Mint unbxd	Good cond	Fair cond
green leaf decals on doors, white lettered decals on sides outlined in black, metal wheels, crimpled axle, OO scale, 75:1. Made 1957	£25	£10	£7	£5
1b Silver grill, grey plastic wheels, crimpled or round axle	£25	£10	£7	£5
1c Deep and light red, silver grill, light grey plastic wheels, round axle. Made 1961/62	£25	£10	£7	£5

No 47 Second issue Lyons Maid ice-cream mobile shop Metallic light blue, silver grill, headlights and bumpers, white interior and seats, man holding ice cream, movable, black base, decals on top and sides, black plastic wheels, knobby treads, length 64 mm. 75:1, OO scale. Made 1963 £25 £10 £6 £3

2a Metallic dark blue, cream or white interior, square decal on top, 3 colour decal on sides, man has legs showing, black plastic wheels, knobby treads £25 £10 £6 £3

	Mint boxed	Mint unbxd	Good cond	Fair cond
2b Flaw: man missing and no decals. Very rare	£250	£100		
2c Dark blue body, white or cream interior, decal on top and 3 colour decal at sides, man has legs showing, black plastic wheels, fine treads	£15	£10	£5	£3
2d Dark or light blue body, decal on top and 3 colour decal on sides, man has no legs showing, blank plastic wheels, fine treads	£15	£10	£5	£3
2e Dark blue, white interior, square decal on top, white decal at sides, man with or without legs showing, black plastic wheels, fine treads, colours varying	£15	£10	£5	£3
2f Medium or light blue, round decal at top, man with or without legs showing, black plastic wheels, fine treads. Made 1966	£15	£10	£6	£3
2g Light blue, off white interior, no decals, no silver on bumpers, grill or headlights. Flaw	£500	£200		

No 47 Third issue ice cream mobile shop Cream body, square decal on top, 3 colour decal on sides, man with legs showing, silver headlights and bumpers, length 64 mm, 85:1. Made 1967 £20 £10 £7 £5

3a Cream body, round decal on top, 3 colour decal on sides £20 £10 £5 £3

3b Cream, round decal on top and white decal at sides, man with or without legs showing £25 £10 £7 £5

3c Dark, medium and light blue body, cream or blue interior on van, dark blue interior on cab, red and white decal on sides, also orange and white decal, man with or without legs showing, very rare, last model to be made. Made 1968 for special order £100 £50 £10 £5

Flaw: no paint on body, off white interior £1000 £250 £25 £10

No 47 Fourth issue Daf tipper container truck Silver blue cab and chassis, orange box, light grey roof, green tinted windows, red grill and quarter base, six black plastic wheels, length 76 mm, 94:1. Made 1969 £7 £5 £3 £2

4a Aqua cab and chassis, light grey roof £5 £4 £2 £1

4b Silver cab and chassis, light grey roof £5 £4 £2 £1

47, Brooke Bond tea van. 48, Meteor sports boat and trailer

48 THE GOLDEN YEARS OF LESNEY

	Mint boxed	Mint unbxd	Good cond	Fair cond
4c Silver cab and chassis, light or dark grey roof, tow slot	£5	£4	£2	£1
4d Silver cab and chassis, dark grey roof, tow slot, superfast wheels. Made 1970	£6	£4	£2	£1
4e Silver cab and chassis, dark grey roof, tow slot, patent app on base, superfast wheels	£5	£4	£2	£1
Flaw: no paint on box, off white roof, superfast wheels, 1971 series	£50	£25		
4f Silver cab and silver grey roof, orange box. Made 1972	£10	£7	£5	£3
No 47 Fifth issue beach hopper Blue with pink pebble dash effect, yellow sun decal on bonnet, fawn plastic driver, orange seats, large wide superfast wheels, length 72 mm, 49:1	£2	£1	50p	25p
5a Dark blue, darker pink pebbledash effect, bright yellow sun decal, wide superfast wheels. Made 1975	£2	£1	50p	25p
No 47 Sixth issue beach hopper Deep purple, pink pebble dash, bright yellow sun decal on bonnet, fawn plastic driver, wide superfast wheels, length 66 mm, 47:1. Made 1976				
No 48 First issue Meteor sports boat on trailer Metal boat, tan deck, blue hull, black metal trailer with tow bar, small hole in boat fits trailer, small round circle at end of tow bar, braces under front axle. Made 1957	£7	£5	£2	£1
1a Metal wheels, crimpled axle, trailer 1/8" longer with flat front tow bar and braces under front axle	£5	£3	£2	£1
No 48 Second issue Meteor sports boat and trailer Very dark blue, dark tan deck, blue hull, black trailer, no braces under front axle, length 67 mm, 75:1, OO scale, metal wheels, crimpled axle. Made 1960	£5	£4	£2	£1
2a Grey plastic wheels, round or crimpled axle	£5	£4	£2	£1

	Mint boxed	Mint unbxd	Good cond	Fair cond
2b Red, white boat, gold engine attached to boat with blue trailer, various shades, length 79 mm, OO scale. Made 1961	£7	£5	£2	£1
2c Red boat, white deck, dark blue trailer, golden motor attached, silver plastic wheels, round axle. Made 1962	£7	£5	£2	£1
2d White boat, deep, medium or light red deck, gold motor, blue trailer. Made 1963-65	£7	£5	£2	£1
No 48 Third issue Sports boat and trailer Brilliant white boat, dark brown deck, gold motor, black plastic wheels, support bar behind hole, length 79 mm, 84:1, HO scale	£5	£4	£2	£1
3a Red deck, light cream hull, dark blue trailer, black tow bar, gold motor	£5	£4	£2	£1
3b Flaw: no paint on boat and gold missing off motor, dark blue trailer	£50	£25		
No 48 Fourth issue dumper truck Deep red body and tipper, green tinted windows, silver plastic grill, base, six black plastic wheels, length 76 mm, 85:1. Made 1967	£7	£5	£2	£1
4a Black plastic wheels with tow slot	£5	£5	£2	£1
Flaw: missing paint from tipper, no windows in cab, five plastic wheels. Made 1969	£100	£50		
4b Orange and yellow dumper, blue metallic cab, silver grill, headlights and tow bar, superfast wheels. Made 1970	£5	£4	£2	£1
4c Light blue cab, bright yellow tipper. Made 1971	£5	£4	£2	£1
4d Yellow dumper, silver metallic cab and chassis, silver plastic wheels. Rare model made 1972	£50	£25		
No 48 Fifth issue Pied Piper Deep purple body, silver plastic motor and exhausts, number 8 decal on roof with stars and red, white and blue stripes, number in yellow, large wide wheels at rear and smaller superfast wheels at front, length 57 mm, 64:1. Made 1973	£2	£1	50p	25p
5a Light blue, otherwise details as above. Made 1974	£2	£1	50p	25p

THE GOLDEN YEARS OF LESNEY 49

	Mint boxed	Mint unbxd	Good cond	Fair cond
5b Medium blue, otherwise details as above. Made 1975	£1	50p	25p	10p
5c Deep blue, length 75 mm, 64:1. Made 1976				

No 49 First issue army personnel carrier Olive green, white circle decal with white star in centre on bonnet, three axles, half base from under bumper to rear of cab, 2 metal wheels, four metal rollers, crimpled axle, 60:1. Made 1958
2 knobby grey plastic wheels, four metal rollers and crimpled axle on second issue made 1959

	Mint boxed	Mint unbxd	Good cond	Fair cond
	£10	£7	£5	£3
1a Two knobby grey plastic wheels, four metal rollers, crimpled or round axles	£10	£7	£5	£3
1b Two knobby light grey plastic wheels, four metal rollers, round axle	£7	£5	£2	£1
1c Two knobby grey plastic wheels, four grey plastic rollers, round axle	£25	£10	£7	£5
1d Two fine large grey plastic wheels, 4 large silver plastic rollers, round axle	£7	£5	£2	£1
1e Two black plastic wheels, four large black plastic rollers, round axle	£25	£10	£7	£5
1f Two large or two small black plastic wheels, four small black plastic rollers, rivet behind front axle	£7	£5	£2	£1
1g Light green body, otherwise details as above, 64 mm, 60:1. Made 1966	£7	£5	£2	£1

No 49 Second issue Unimog Tan cab and tipper, green chassis, green tinted windows, orange plastic wheels, black plastic tyres, length 62 mm, 68:1

	Mint boxed	Mint unbxd	Good cond	Fair cond
	£7	£5	£3	£2
2a Yellow hubs, spare tyre on left side under cab and bed of truck, 4 black plastic tyres, 2 axles and tow hook, brace in front of front axle and rear of rear axle	£7	£5	£3	£2

No 49 Third issue Unimog Blue cab and body, red chassis, red front bumper and tow hook, front brace only, green tinted windows, orange hubs, spare wheel on left side under cab on bed of truck, 62 mm, 68:1. Made 1968

	Mint boxed	Mint unbxd	Good cond	Fair cond
	£7	£5	£3	£2

3a Blue cab and red body, red chassis, front bumper and tow hook, front brace only with brace

	Mint boxed	Mint unbxd	Good cond	Fair cond
under axle wider. Made 1969	£7	£5	£2	£1
3b Superfast balloon tyres. Made 1970	£6	£4	£2	£1
3c Dark red chassis, balloon tyres	£5	£4	£2	£1
3d Metallic light blue cab and chassis, balloon tyres	£5	£4	£2	£1
3e Metallic light blue body, blue grill, balloon tyres with superfast wheels	£5	£4	£2	£1
3f Lighter blue or very bright blue body, large superfast wheels, thin tread. Made 1972	£5	£4	£2	£1

No 49 Fourth issue Chop Suey Red chassis, silver forks, engine and frame, dark red seat, yellow head, black wide wheels, length 70 mm. Made 1973

	Mint boxed	Mint unbxd	Good cond	Fair cond
	£1	75p	50p	25p
4a Red forks, otherwise details as above. Made 1974	£1	75p	50p	25p
4b Deep red frame, darker red seat. Made 1975	£1	75p	50p	25p
4c Dark red forks, deep purple seat, length 72 mm, 50:1. Made 1976				

No 50 First issue Commer pick-up truck Tan body, black base, silver grill, headlights and bumpers, metal wheels and crimpled axle, 64:1, mark 8. Made 1958

	Mint boxed	Mint unbxd	Good cond	Fair cond
	£5	£4	£2	£1
1a Tan body, grey plastic wheels, crimpled axle	£5	£4	£2	£1
1b Light tan or ordinary tan, grey plastic wheels, crimpled or round axles	£5	£4	£2	£1
1c Dark tan, grey plastic wheels, round axle, some models with rear tan bumper	£5	£4	£2	£1
Flaw: Paint completely missing from body	£100	£50		
1d Dark tan, silver plastic wheels, knobby treads	£5	£4	£2	£1
1e Bright orange, yellow bumpers and grill, white trim. Made 1962	£10	£5	£2	£1

No 50 Second issue Commer pick-up truck Red body and chassis, white cab roof and truck interior, silver grill, bumpers and headlights, grey plastic wheels, knobby treads, length 67 mm, 64:1, Made 1963

Mint boxed	Mint unbxd	Good cond	Fair cond
£10	£5	£2	£1

50   THE GOLDEN YEARS OF LESNEY

	Mint boxed	Mint unbxd	Good cond	Fair cond
2a Red and grey, red bumpers, grey plastic wheels, knobby treads	£5	£4	£2	£1
2b Red and grey, silver or black plastic wheels, knobby treads	£5	£4	£2	£1
2c Red and grey, or red and white cab, dark red body, black plastic wheels, knobby treads. Made 1963	£5	£4	£2	£1
Flaw: paint completely missing off lower part of truck and only very thin off white paint on top of cab	£50	£25		
No 50 Third issue John Deere tractor Green, yellow hubs and yellow or black steering wheel, green tow hook, length 54 mm, 61:1. Made 1964	£10	£5	£2	£1
3a Grey or black plastic tyres	£7	£5	£2	£1
3b Green, silver grill, orange hubs, white steering wheel and tow hook. Made 1965	£10	£7	£5	£3
3c Light green, yellow hubs, blue grill, yellow steering wheel, length 54 mm. Made 1966	£5	£4	£2	£1
3d Green, black or yellow steering wheel, light and dark yellow hubs, yellow hook. Made 1967/8	£6	£4	£2	£1
4a Dark green, white grill, tinted or clear canopy	£5	£4	£2	£1
4b White or silver grill, textured bed. Made 1969	£5	£4	£2	£1
4c Light metallic green, medium and light green, silver grill, textured bed, superfast wheels. Made 1970	£4	£3	£2	£1
No 50 Fifth issue kennel truck Light pea green, superfast wheels, length 70 mm, 67:1. Made 1973	£4	£3	£2	£1
5a Light or dark yellow base, silver grill, textured bed, wide superfast racing wheels. Factory had run short of normal superfast, and therefore a few of these models are worth a considerable sum	£50	£25	£10	£5

49, Army personnel carrier. 50, Commer pickup

51, Albion Chieftain. 52, Maserati 4CLT

THE GOLDEN YEARS OF LESNEY 51

	Mint boxed	Mint unbxd	Good cond	Fair cond
No 50 Sixth issue articulated truck Fawn cab and chassis, blue box with orange flashes on side, black grill, orange tinted windows, four superfast wheels on cab and two on rear of chassis, length 76 mm, 90:1. Made 1974	50p	25p		
6b Very light yellow cab and chassis, very light blue wagon body. Made 1976				
No 51 First issue Albion Chieftain Yellow, tan load, silver grill and headlights, small round decal on doors, Portland Cement decal in two lines, metal wheels, crimpled axle, 150:1	£10	£5	£2	£1
1a Yellow, light tan load, Portland cement decal in two lines	£10	£5	£2	£1
1b Yellow, light tan load, blue circle Portland cement decal in two lines, metal wheels, crimpled axle. Made 1959	£10	£5	£2	£1
1c Yellow orange, light tan load, blue circle Portland cement decal in two lines, metal wheels, crimpled axle. Made 1960	£7	£5	£2	£1
1d Yellow orange, tan and light tan load, or dark tan load, large or small grey plastic wheels, crimpled or round axle	£7	£5	£2	£1
1e Yellow, light tan, tan or dark tan load, silver plastic wheels, round axle	£5	£4	£2	£1
1f Yellow orange, tan, light tan and dark tan loads, small silver plastic wheels	£5	£4	£2	£1
1g Yellow, dark tan load, very small black plastic wheels, knobby treads. Made 1963	£7	£5	£2	£1
No 51 Second issue John Deering trailer Green, yellow or black hubs, brown barrels, length 67 mm, 61:1. Invented by great mechanical engineer, John Deere. Made 1964	£2	£1	75p	50p
2a Green metal trailer and chassis, orange hubs and barrels, grey or black plastic tyres. Varying colours but prices fairly stable up to last issue in 1968. Made 1965	£2	£1	75p	50p

	Mint boxed	Mint unbxd	Good cond	Fair cond
No 51 Third issue 8 wheel tipper Orange cab and chassis, silver tipper with decals, eight plastic wheels, silver grill, Douglas decal, length 76 mm, 92:1. Made 1969	£10	£5	£3	£2
3a Orange cab and chassis, silver grill, orange Douglas sticker decal	£5	£4	£3	£2
3b Orange cab and chassis, tow slot, orange Douglas sticker decal	£5	£4	£3	£2
3c Deep, medium and light yellow cab and chassis, silver base and grill, orange Douglas sticker decal. Made 1969	£7	£5	£3	£2
No 51 Fourth issue 8 wheel tipper Orange cab, blue tinted windows, yellow pointer decals on sides, silver grill, headlights and bumpers, eight black plastic wheels. Made 1970	£7	£5	£3	£2
4a Light and medium yellow body, pointer decals, superfast wheels. Made 1971/2	£5	£4	£2	£1
Flaw: Paint completely missing off cab and forward part of chassis, no decals, silver tipper badly made, shorter at one side than the other. Made 1972	£250	£100		
No 51 Fifth issue Citroen SM Deep rose pink, silver grill, hubs, headlights and bumpers, white or cream interior, opening doors, cream or white tow hook, superfast wheels, length 58 mm 63:1. Made 1973	£2	£1	50p	25p
5a Light pink, brown interior, white plastic grill, headlights and tow bar and bumpers. Made 1974	£1	75p	50p	25p
No 51 Sixth issue Citroen SM Light, medium and dark pink, silver plastic grill, headlights and bumpers, tow hook, wide superfast wheels, length 58 mm, 63:1. Made 1975	£1	75p	50p	25p
Flaw: Paint completely missing from top of bonnet and drivers side door, no steering wheel	£50	£25		
6a Light, medium and dark blue body, white interior, orange decals on roof, bonnet bordered				

THE GOLDEN YEARS OF LESNEY

	Mint boxed	Mint unbxd	Good cond	Fair cond
with white and blue flashes, number 8 decal on roof. Made 1976				
No 52 First issue Maserati 4CLT Red, cream driver, silver grill, black base, number 52 decal on sides, knobby black plastic wheels, crimpled axle, 63:1, unique first racing car model brought on racing scene in 1948. Made 1958	£10	£5	£2	£1
1a Red and silver, exhaust on sides, gas cap behind driver, knobby black plastic wheels, crimpled axle	£7	£5	£2	£1
1b Red, white driver, silver exhausts and gas cap, number 52 decal in black numbers inside white circle, knobby black plastic wheels, crimpled axle	£7	£5	£2	£1
1c Red body, red gas cap, knobby black plastic wheels, round axle	£7	£5	£3	£2
1d Pink body, yellow driver, wire wheels, silver grill, black balloon tyres. Made 1963	£10	£7	£5	£3
No 52 Second issue Maserati 4CLT Yellow body, cream driver, silver grill, number 52 decal, wire wheels, yellow silver grill, black plastic tyres, length 64 mm, 63:1. Made 1964	£10	£7	£5	£3
2a Yellow body, silver grill, cream driver with arm patch on left shoulder, wire wheels. Made 1965	£7	£5	£3	£2
Flaw: no paint on body, only one wire wheel, no grill	£100	£50		
Flaw: with decal 25	£100	£50		
No 52 Third issue BRM racing car Blue, white plastic driver and steering wheel, yellow or orange hubs, metal engine and base plate, black plastic tyres, length 70 mm, 54:1. Made 1966	£5	£4	£2	£1
3a Blue, number 5 decals on both sides and hood, in black numbers inside white circle	£5	£4	£2	£1
3b Decals on hood, sticker decals on sides. Made 1967	£25	£10	£7	£5
3c Sticker decals on hood and sides	£4	£3	£2	£1
3d Red number 5 decals on both sides and hood. Very rare model	£100	£50	£25	£10

	Mint boxed	Mint unbxd	Good cond	Fair cond
3e Red body, red sticker decals on hood and sticker decals on side, only few made to complete order in 1969	£50	£25	£7	£5
Flaw: Model with superfast wheels.	£50	£25		
Above models come in various shades				
No 52 Fourth issue Dodge charger Mk 3 Rose pink, white interior, clear windshield, green base plate, wide superfast racing wheels, length 76 mm, 62:1. Made 1970	£5	£3	£2	£1
4a Deep red, superfast wheels. Made 1971	£5	£4	£2	£1
4b Light pink, sticker decals on sides, number 5 on roof. Made 1972	£5	£4	£3	£1
4c Dark red, green base plate, decals on sides, number 5 on roof. Made 1972	£4	£3	£2	£1
4d Dark purple body, red green base plate, sticker decals on roof and sides or without any decals. Made 1973	£2	£1	75p	50p
4e Deep gold body, no decals. Made 1974	£1	75p	50p	25p
4f Dark gold, black interior, silver base plate. Made 1975	£1	50p	25p	10p
4g Metallic green, length 75 mm, 62:1. Made 1976				
No 53 First issue Aston Martin DB2 saloon Silver green, silver headlights and bumpers, red tail lights, metal wheels, knobby treads, crimpled axle, OO scale, 68:1. Made 1959	£10	£4	£3	£2
1a Silver green, silver grill, grey plastic wheels, crimpled axle	£7	£5	£2	£1
1b Light metallic green, silver grill, grey plastic wheels, crimpled or round axle. Made 1962	£5	£4	£2	£1
1c Metallic red, grey plastic wheels, knobby treads, rare, very few made	£50	£25	£10	£7
1d Metallic red, black plastic wheels, knobby treads	£50	£25	£10	£7
No 53 Second issue Mercedes Benz 220S Metallic				

THE GOLDEN YEARS OF LESNEY 53

	Mint boxed	Mint unbxd	Good cond	Fair cond
red, silver grill, white interior, silver bumpers and headlights, length 70 mm, 73:1, OO scale. Made 1963	£10	£7	£5	£3
2a Maroon, grey plastic wheels	£7	£5	£2	£1
2b Maroon, silver plastic wheels	£5	£4	£3	£2
2c Red, medium red and light red, grey or silver plastic wheels	£7	£5	£2	£1
2d Red, red rear bumper, black plastic wheels, fine treads	£10	£7	£5	£3
2e Red, red rear bumper, very dark red doors, black plastic wheels	£7	£5	£3	£2
2f Red bronze, gold interior, blue grill and headlights, silver bumpers, red tail lights, black plastic wheels, length 70 mm, 73:1, OO scale. Made 1966 for International year	£25	£15	£10	£5
2g Deep chocolate, gold interior, blue headlights, grill and bumpers, blue black plastic wheels. Made 1967	£25	£15	£10	£5

No 53 Third issue Ford Zodiac Mk IV Silver blue, off white interior, tow slot, metal grill, bumpers and base plate, hood lifts up showing silver plastic motor with spare wheel, black plastic wheels, length 70 mm, 66-1 | £10 | £5 | £3 | £2 |
| **3a** Dark and medium blue. Made 1968 | £10 | £5 | £3 | £2 |

No 53 Fourth issue Ford Zodiac Mk IV Metallic rich green or dark green, silver grill, bumpers and headlights, orange flashers, black base, white interior, length 70 mm, 66:1. Made 1970 | £10 | £5 | £2 | £1 |
4a Light metallic green, superfast wheels. Made 1971	£7	£5	£2	£1
4b Light, medium and dark metallic green, wide racing wheels, nice model. Made 1972	£7	£5	£3	£2
Flaw: two racing wheels are placed at rear and two thin superfast wheels at front, missing grill	£25	£10		

No 53 Fifth issue Tanzara Orange red, opening rear cover to show silver plastic motor, green tinted windows, silver grill, wide black plastic superfast wheels, length 76 mm, 60:1. Made 1973 | £2 | £1 | 75p | 50p |

	Mint boxed	Mint unbxd	Good cond	Fair cond
5a Dark orange red, superfast wheels, made 1974	£1	75p	50p	25p
5b Red, superfast wheels	£1	50p	25p	10p
5c Flaw: Motor missing and body badly twisted on vehicle	£10	£5		

No 53 Sixth issue Tanzara White, red, white and blue decals stars and stripes decals on roof and bonnet, orange interior, orange or black base, silver grill, bumper and motor, wide superfast wheels. Made 1976

No 54 First issue Saracen personnel carrier Olive green, rotating gun turret, 3 axles, 6 black plastic wheels, length 57 mm, 86:1, HO scale. Made 1959 | £10 | £5 | £3 | £2 |
1a Knobby black plastic wheels, crimpled axle. Made 1960	£10	£5	£3	£1
1b Knobby black plastic wheels, round axle	£10	£3	£2	£1
1c Black plastic wheels, fine treads, round axle	£5	£3	£2	£1

53, Aston Martin DB Saloon. 54, Saracen carrier

54 THE GOLDEN YEARS OF LESNEY

	Mint boxed	Mint unbxd	Good cond	Fair cond

Flaw: back wheels are very much smaller than the others — £100 £25

No 54 Second issue Cadillac ambulance White, silver grill and headlights, white interior, red dome lights, red cross decals on sides, length 73 mm, 87:1, HO scale. Made 1966 — £25 £10 £5 £3

No 54 Third issue Cadillac ambulance White, blue tinted windows, small cross decals on doors, black plastic wheels, length 73 mm, 87:1, Made 1967 — £10 £5 £3 £2

3a Large cross decals, black plastic wheels — £7 £5 £3 £2

3b Small cross sticker decals, black plastic wheels — £7 £5 £2 £1

3c Flaw: no decals, two superfast wheels, two non-superfast wheels, box unprinted on front, very rare — £1,000 £100

No 54 Fourth issue Cadillac ambulance Cream body, silver grill, headlights, bumpers, superfast wheels. Made 1970 — £7 £5 £2 £1

Flaw: missing red cross on box of one of the above issues, other faults in model itself, bad paintwork and thin tyres — £1,000 £100

No 54 Fifth issue Ford Capri Orange red, off white interior, metal grill and base, plastic tow hook, opening bonnet reveals silver plastic motor, length 73 mm, 59:1. Made 1971 — £3 £2 £1 50p

5a Pink, dark plastic motor, black bonnet, otherwise as above. Made 1972 — £2 £1 50p 25p

5b Deep pink, black bonnet, wide racing wheels. Made 1973 — £1 75p 25p 10p

5c Rose pink, white interior, metallic pink opening bonnet, plastic headlights, bumpers and grill. Made 1975 — £1 50p 25p 10p

No 54 Sixth issue Ford Capri Metallic red, opening bonnet, silver motor, headlights and grill, silver bumper. Made 1976

55, DUKW. 56, London trolley bus

	Mint boxed	Mint unbxd	Good cond	Fair cond

No 55 First issue DUKW Olive green, black base, 3 axles and 6 metal wheels, OO scale, 70:1, crimpled axle, amphibious craft. Made 1959 — £7 £5 £3 £2

1a Grey plastic front wheels, four metal rear wheels — £50 £25 £10 £5

1b Grey plastic wheels, crimpled axle, length 73 mm, 70:1, OO scale. Made 1962 — £7 £5 £3 £2

No 55 Second issue police car Silver grill, headlights and bumpers, red dome light with white decal on hood, black lettering and red white

THE GOLDEN YEARS OF LESNEY 55

	Mint boxed	Mint unbxd	Good cond	Fair cond
and blue shield, shield decals on doors, no occupants, length 67 mm, 80:1, OO scale, Fairlane model. Made 1963	£10	£7	£5	£2
2a Dark or light blue, black plastic wheels, round axle, knobby treads	£7	£5	£2	£1
2b Light or medium blue, black plastic wheels, fine treads	£5	£4	£2	£1
2c Light blue, black plastic wheels, fine treads, blue rear bumper	£7	£5	£2	£1
Flaw: missing decals on vehicle, only half decals on doors. Bad factory flaw	£250	£50		
No 55 Third issue police car White, off white interior, tow hook, red dome light, white decals on hood and door, red yellow and blue shields, one police driver, 55/59 stamped on base, length 73 mm, 73:1, OO scale, Ford Galaxy model. Made 1966	£10	£5	£2	£1
3a Blue dome light, silver metal hubs, black plastic wheels. Made 1966	£7	£5	£2	£1
3b Blue dome light and red dome light, superfast wheels. Made 1970	£5	£4	£2	£1
3c Red dome light, superfast wheels, no separation between mercury platform and base	£10	£5	£2	£1
Flaw: dark cream on one side, white colour on driver's side, only half a driver	£100	£50		
No 55 Fourth issue Mercury police car White, off white interior, no driver, red double dome lights, blue double dome lights, base with tow slot, wide racing wheels. Made 1971	£10	£5	£2	£1
4a Single dome red light. Made 1972	£5	£4	£2	£1
4b Double red dome lights, no decal on doors, red, yellow and blue Police decal on bonnet. Made 1973/74	£2	£1	50p	25p
4c Off white, double red dome lights, cheapened version of above model. Made 1975	£1	50p	25p	10p
No 55 Fifth issue Hellraiser White, red interior, silver plastic motor in rear, streamlined body, star decalled				

	Mint boxed	Mint unbxd	Good cond	Fair cond
on bottom rim, red, white and blue striped decals with three stars on blue square, wide superfast racing wheels, length 75 mm, 49:1. Made 1976				
No 56 First issue London trolley bus Red, two trolley poles on roof, large decals on each side, 'Drink Peardrax', letters in black and red, small decal on left side, front and back in white letters on black background, 6 metal wheels, 3 crimpled axles, 137:1. Made 1959/60.	£50	£25	£10	£5
1a Black trolley poles on roof, white background decal, metal wheels, crimpled axles	£50	£25	£10	£5
1b Red trolley poles, grey outline on Peardrax advert with deep orange letters, metal wheels, crimpled axles	£100	£50	£25	£10
1c Deeper red trolley bus, Peardrax advert with orange letters and small letters in purple. Made 1963	£25	£15	£10	£5
1d Red trolley poles, white or yellow background decals, grey plastic wheels, round or crimpled axles. Made 1964-5.	£50	£25	£10	£5
1e Red trolley poles, white background decals, black plastic wheels, round axle	£50	£25	£10	£5
Flaw: decals at uneven angle, Peardrax, no decals on front or sides at rear of trolley	£250	£100		
No 56 Second issue Fiat 1500 Green, red interior, red steering wheel, silver grill, headlights and front bumper, red tow hook, black plastic base plate, brown plastic luggage on roof rack, length 67 mm, 64:1. Made 1966	£7	£3	£2	£1
2a Green, dark brown luggage, or luggage reversed, black plastic wheels. Made in 1967	£4	£3	£2	£1
2b Green, tan luggage, bumpers not painted silver, black plastic wheels,	£7	£5	£2	£1
2c Green, tan luggage reversed, green bumpers, black plastic wheels. Made 1969	£5	£4	£3	£1
2d Red, tan luggage, red bumpers, black plastic wheels. Very rare model. Made 1969 for special order	£50	£25	£10	£5

THE GOLDEN YEARS OF LESNEY

	Mint boxed	Mint unbxd	Good cond	Fair cond

No 56 Third issue BMC 1800 Pininfarina metallic gold body, silver metal base, headlights and bumpers, off white interior, clear windows, front doors open, tow hole, superfast wheels, length 70 mm, 64:1, Made 1970 — £5 £3 £2 £1

3a Dark metallic gold. Made 1971 — £3 £2 £1 50p

3b Bright metallic orange. Made 1972 — £1 75p 50p 25p

3c Very pale orange and bright metallic orange. Made 1973/4 — £1 50p 25p 15p

No 56 Fourth issue Hi tailer Off white body, blue driver, silver motor, red, white and blue decals with black lettering MB and letter 5 in white circle with dark orange ring surround, length 72 mm, 58:1. Made 1975/6 — 50p 25p 10p

4a Dark blue driver, very black lettering on decals, length 76 mm. Made 1976

No 57 First issue Wolseley 1500 Pale yellow green, gold or silver grill, headlights and bumpers, red tail lights, grey plastic wheels, crimpled axle. Made 1959 — £7 £5 £2 £1

1a Pale yellow green, grey plastic wheels, round axle. Made 1960 — £5 £3 £2 £1

No 57 Second issue Chevrolet Impala Two tone, pale blue roof with metallic purple body, blue tinted windows, silver grill, headlights and bumpers, red tail lights, metal base, tow hook, grey plastic wheels, length 80:1, 66 mm, OO scale. Made 1961 — £10 £5 £3 £2

2a Metallic blue, pale blue roof, black base, blue rear bumper, grey plastic wheels. Made 1962 — £7 £5 £3 £2

2b Metallic blue, white roof, dark gold trim, blue base, length 70 mm, 80:1, OO scale. Made 1963 — £7 £5 £3 £2

2c Black or blue base, silver bumpers, silver plastic wheels, knobby treads — £5 £3 £2 £1

2d Blue base, grill and bumpers, silver plastic wheels — £5 £3 £2 £1

	Mint boxed	Mint unbxd	Good cond	Fair cond

2e Black base, silver grill, front bumper, red tail lights — £5 £4 £2 £1

No 57 Third issue Chevrolet Impala Dark blue, cream roof, gold line trim, silver headlights, grill and bumpers, silver plastic wheels, length 70 mm, 80:1, OO scale. Made 1964 — £10 £5 £2 £1

3a Black base, silver grill, blue bumpers and tail lights — £5 £4 £2 £1

3b Blue, black base, no silver trim — £15 £10 £7 £5

3c Light blue streamlined body, off white interior, prominent gold trim on fins and front headlights, bumpers and grill, green tinted windows, bright cream roof, gold trim all over body, blue tail lights, black wheels, knobby treads. Made 1976 — £10 £5 £3 £2

Flaw: paint completely missing from roof, lower body paint badly flecked — £100 £50

No 57 Fourth issue Land Rover fire truck Red, blue tinted windows, blue dome light, blue plastic ladder on roof, red grill, headlights and bumpers, Kent Fire Brigade decals in yellow and red shield with white crest on side of doors, length 64 mm, 71:1, OO scale. Black plastic wheels, Made 1966 — £15 £10 £5 £3

4a Decals as above, grey plastic wheels, fine treads — £7 £5 £3 £2

4b Decals, red head lights, black plastic wheels, fine treads — £7 £5 £3 £2

4c Decals as above, blue dome, white plastic ladder on roof, silver headlights, front bumper, black plastic wheels — £10 £5 £3 £2

4d Red headlights, sticker decals, black plastic wheels, fine treads. Made 1969 — £5 £4 £2 £1

Flaw: Decals missing from doors, grey plastic rear wheels, black superfast front type — £25 £10

No 57 Fifth issue Eccles caravan Fawn, orange removable roof, green interior, tow bar, dark

THE GOLDEN YEARS OF LESNEY

	Mint boxed	Mint unbxd	Good cond	Fair cond

brown stripe on sides, long overhang, superfast wheels, length 83 mm, 76:1. Made 1970 — £7 £5 £3 £2

5a Dark brown stripe on sides, short overhang — £5 £3 £2 £1

No 57 Sixth issue Eccles caravan White, orange roof, light brown strip on side, flowered decal on side, superfast wheels. Made 1971 — £7 £5 £3 £2

6a Dark maroon stripe on sides, long or short overhang — £5 £3 £2 £1

6b Maroon stripe on sides, flower at rear, long or short overhang — £5 £3 £2 £1

6c Dark fawn, orange stripe on sides, orange flowered decal, dark interior, inferior roof. Made 1973 — £4 £3 £2 £1

Please note the rear window comes with an overhang that is either bevelled on the ends or is sharply rounded to the sides on most of the above models. This gives you a long or short overhang and the long overhang model has two braces on the tow bar near the body; the short overhang model braces are a bit smaller; these models also come with various lettering. The letter C and the letter R come in circles or inside ovals, or an oval and circle both together. If you combine all variations together you could add hundreds of variations for this one particular model alone.

No 57 Seventh issue Wild Life Truck Mustard body, red interior and red plastic base, elephants head decal on bonnet, lion in back, clear plastic cover, lion revolves on a circular drum, length 72 mm, 67:1. Made 1974 — £1 75p 25p 10p

7a Dark brown body, yellow lion, bright orange decal with white elephants head on bonnet, silver grill, headlights and bumpers. Made 1975 — £1 50p 25p 10p

7b Bright yellow, silver plastic bumper, grill and headlights, red plastic base, dark orange lion in rear, white elephant decal on orange background, with 'Ranger' in black, length 71 mm. Made 1976

No 58 First issue BEA coach Dark blue, silver headlights and shield ornament on front, two decals on side with British European Airways in white, metal wheels, length 139:1. Made 1959 — £100 £50 £10 £5

1a Dark blue, silver headlights and shield ornament, two decals either clear with British European Airways in white, or a white decal with red box inside white BEA and white background with black letters, British European Airways on sides, grey plastic wheels. Made 1960 — £100 £50 £10 £5

No 58 Second issue BEA coach White decals with BEA in front, grey plastic wheels, crimpled axle, length 72 mm. Made 1961 — £25 £15 £10 £5

2a Decals BEA in front, light grey plastic wheels, round axle — £25 £15 £10 £5

2b Blue or bright blue body, BEA at rear, dark grey plastic wheels, round axle — £25 £15 £10 £5

2c Very dark blue body, BEA in front and also at rear, silver plastic wheels, round axles. Made 1962 — £50 £25 £10 £5

2d White decals, BEA in front or at rear, black plastic wheels, knobby treads. Made in 1962 — £25 £15 £10 £5

Flaw: Missing decals completely from behind and decals upside-down on front of coach — £150 £50 £25 £10

Please note that there are possible chances that BEA sign comes in all models in front and rear of decals

No 58 Third issue Drott excavator Red, movable shovel, four plastic rollers, green rubber treads, length 67 mm, 60:1. Made 1963 — £15 £10 £5 £3

3a Red, silver motor and base, silver plastic rollers — £7 £5 £3 £2

3b Red, silver motor and base, black plastic rollers — £7 £5 £3 £2

3c Orange, silver motor and base, green plastic rollers, green rubber treads. Made 1964 — £10 £7 £5 £3

3d Orange body, motor and base, black plastic rollers. Made 1965 — £10 £7 £5 £3

3e Orange body, silver grill and motor, extra large shovel in front, black plastic rollers, green rubber

58 THE GOLDEN YEARS OF LESNEY

	Mint boxed	Mint unbxd	Good cond	Fair cond
treads, length 67 mm, 60:1. Made 1966 for International Year	£25	£15	£10	£5
3f Smaller shovel in front, yellow trim all round body. Made 1967	£10	£7	£5	£3

No 58 Fourth issue Daf girder truck Orange body, ribbed bed with four stakes, one in each corner, red plastic girders, fawn tinted windows, red plastic grill and quarter base, black plastic wheels, round axle, length 76 mm, 94:1. Made 1968 — £7 £5 £3 £2

4a Cream body, nothing holding plastic windows to roof, also very bad warp in body, factory flaw — £25 £15 £2 £1

4b Cream body, pin holding plastic windows to roof, tow slot — £5 £4 £2 £1

4c Silver grey body and chassis, red load, green tinted windows, red grill and tow slot, blue black plastic wheels, black decals on front, licence plate. Made 1969 — £10 £5 £3 £2

4d White body, red grill and base, white petrol tank and licence number on front mudguard, silver headlights. Made 1970 — £5 £4 £2 £1

4e Metallic gold body, red grill and orange plastic load. Made 1971 — £4 £3 £2 £1

4f Metallic golden green body, other details as above. Made 1972 — £4 £3 £2 £1

Flaw: shortened load which is apparent by half the girders being shorter than others as materials must have run short at the factory as was sometimes the case. Missing grill and metallic paint very poor quality — £25 £10 £5 £3

No 58 Fifth issue Woosh-n-push Yellow body, silver grill, number 2 in large letter on back window, wide superfast racing wheels, red seats and base, length 76 mm. Made 1973 — £1 75p 20p 10p

	Mint boxed	Mint unbxd	Good cond	Fair cond
5a Bright yellow body, brighter orange seats and base, white plastic grill, black letter 2 on rear. Made 1974	£1	50p	20p	10p
5b Bright lemon body, orange interior and base, silver plastic hubs and grill, number 2 in blue. Made 1975	75p	50p	20p	10p

No 58 Sixth issue Woosh-n-push Bright metallic red, cream seats and base, red, white and blue decal on top at rear, length 77 mm, 60:1. Made 1976

No 59 First issue Ford Thames 5 cwt van "Singer" Light green, silver grill, headlights and bumpers, orange decals on sides, outlined with golden yellow, 'Singer Sewing Machines' decals on doors inside large S, grey plastic wheels, crimpled axle, 64:1. Very rare model. Made 1959 — £75 £50 £15 £10

1a Light green, two metal lines on door with S in between, grey plastic wheels, round axle — £25 £15 £7 £5

1b Light green, silver grill, headlights and bumpers, fully decalled, grey plastic wheels, round axle — £25 £15 £7 £5

1c Light blue, fully decalled. Made 1961 — £25 £15 £10 £5

1d Very light blue green. Made 1962 — £25 £15 £10 £5

1e Dark and medium green, silver headlights, grill and bumpers, fully decalled, silver plastic or large silver plastic wheels, round axle, length 57 mm, 64:1. Made 1963 — £50 £25 £10 £5

1f Kelley dark green, green bumper at rear, black plastic wheels, round axle, knobby treads — £50 £25 £10 £5

Please note that there are various shades of green as far as the vans are concerned, but only an expert can pick out the true colours. There are always variations on the lettering on the sides, especially where the red is concerned

No 59 Second issue Ford Fairlane fire chief's car Red or flame red, white interior, blue dome light,

THE GOLDEN YEARS OF LESNEY 59

	Mint boxed	Mint unbxd	Good cond	Fair cond
silver grill, bumpers and headlights, yellow decal 'Fire Chief' on sides and on hood, length 67mm, HO.OO scale. Made in 1964	£10	£5	£2	£1
2a Deep red, details as above. Made 1965	£7	£5	£2	£1
2b Red, shield decals on doors, fire chief decals on bonnet, blue headlights, grill and bumpers, silver trim, headlights, etc, black plastic wheels, length 73mm. Made 1966	£5	£4	£2	£1
2c Red, sticker decals on hood and doors, number 59 on raised platform. Made 1967-69	£5	£4	£2	£1

No 59 Third issue Mercury fire car Red, off white interior, two occupants, blue dome light, number 59 or 73 on base, Mercury on raised platform, sticker decals in various shades, metal grill, bumpers and base, superfast or wide superfast wheels. Made 1970 — £5 £4 £2 £1

No 59 Fourth issue fire chief car Red, off white interior, blue flash on top, shield decal on doors, fire chief decal on bonnet, white plastic grill, headlights and bumpers, superfast wheels. Made 1971/2/3 — £4 £2 £1 50p

4a Black, red and yellow, gaily coloured decal on doors and bonnet — £1 75p 25p 10p

4b Darker red, deteriorated quality on model. Made 1975 — 50p 25p

No 59 Fifth issue Planet Scout car Rich green tone on top half of body, bright yellow on lower body and base, silver headlights and grill, yellow bumper bars, orange interior or off white interior, orange tinted windows, wide superfast wheels, length 70 mm, 73:1. Made 1976

No 60 First issue Morris J pickup truck Blue body, black base, silver grill, headlights and bumpers, open front window, sides and rear window, red and black lettering on sides, grey plastic wheels, crimpled axle, 75:1. Made 1959 — £10 £5 £2 £1

1a Red and white lettering on sides, grey plastic wheels, round or crimpled axle. Made 1960 — £7 £5 £2 £1

	Mint boxed	Mint unbxd	Good cond	Fair cond
1b Blue rear bumper, grey plastic wheels, round axle, length 56 mm, OO scale. Made 1961	£7	£5	£2	£1
1d Blue rear bumpers, large silver plastic wheels, round axle	£5	£4	£2	£1
1e Blue rear bumpers, black plastic wheels, round axle, fine treads. Made 1963	£5	£4	£2	£1
1f Blue rear bumpers, black plastic wheels, knobby treads, length 57 mm. Made 1964	£7	£5	£2	£1
1g Medium or light blue, no rear window, black plastic wheels, fine treads. Made 1965/66	£7	£5	£3	£2
Flaw: Very short body, unusual shape of cab, no silver trim or decals	£50	£25		

No 60 Second issue site hut truck Deep or light blue cab, chassis with orange hut and green roof, blue tinted windows, silver plastic grill, headlights and base plate, black plastic wheels, length 65 mm, 92:1. Made 1967 — £7 £5 £3 £2

2a Very dark royal blue body and chassis, light tan hut, green roof with fine yellow trim, silver blue bumper, grill and headlights, black plastic wheels with fine tread. Made 1968 — £7 £5 £3 £2

2b Medium blue, tow slot, black plastic. Made 1969 — £7 £5 £3 £2

2c Bright blue body and chassis, silver bumper bar, grill and headlights, bright orange hut with green roof, superfast wheels. Made 1970 — £7 £5 £2 £1

2d Pat App reads upside down on base at rear, tow slot, superfast wheels — £10 £3 £2 £1

2e Light blue body and chassis, cheap plastic hut with green roof, bad deterioration in colour and quality. Made 1971 — £2 £1 50p 25p

No 60 Third issue Lotus super seven Orange yellow body, black seats and base, plastic windscreen, black, red and yellow flame decal on bonnet, superfast wheels, length 52 mm, 51:1. Made 1972 — £2 £1 50p 25p

3a Deep orange, black seats, silver trim, flower decal on hood. Made 1973 — £1 50p 20p 10p

THE GOLDEN YEARS OF LESNEY

	Mint boxed	Mint unbxd	Good cond	Fair cond
3b Lighter orange body, otherwise details as above. Made 1974/5	£1	50p	20p	10p

No 60 Fourth issue Lotus super seven Yellow body, black seats, blue pointed flash decal and red orange design with number 60 in yellow on orange background on hood and on mudguards at rear, silver plastic base, plastic windscreen, length 74mm, 51:1. Made 1976

No 61 First issue Ferret scout car Olive green, tan driver with tan cap, holding tan steering wheel, Matchbox series on black base, 4 grey plastic wheels, knobby treads, 67:1. Madd 1959

	£10	£7	£5	£3
1a Black plastic wheels, crimpled axle, knobby treads	£10	£7	£5	£3

No 61 Second issue Ferret scout car Olive green, black base, yellow driver with yellow cap and yellow steering wheel, spare wheel on side, grey plastic wheels. Made 1961

	£10	£7	£5	£3
2a No Matchbox series on base, black plastic wheels, knobby treads, crimpled axle	£4	£4	£2	£1
2b No Matchbox series on base, black plastic wheels, knobby treads, round axle	£5	£4	£2	£1
2c Matchbox series on base, driver facing the rear, black plastic wheels, round axle, fine treads	£25	£10	£5	£3
2d Driver facing normal way, black plastic wheels, fine treads, round axle	£5	£4	£2	£1

59, Thames 5 cwt Singer van.
60, Morris J pickup

61, Ferret scout car.
62, General service lorry

	Mint boxed	Mint unbxd	Good cond	Fair cond

No 61 Third issue military scout car Lime green, tan driver, tan hat and driving wheel, large plastic wheels, fine treads, black base with Matchbox series

	£7	£5	£2	£1
Flaw: driver with genuine head missing	£50	£25		
3a Very light green, dark tan driver, blue black plastic wheels. Made for International 1966	£25	£15	£7	£5

Please note that over each wheel well you will find a three corner block or empty space. Verified models show over 20 variations with or without blocks for each variation listed above; there are possibilities of over 50 variations on each model not counting the driver which faces either to the front or rear. All in all there are well over 350 variations on this model

No 61 Fourth issue Alvis Stewart White body, yellow plastic removable canopy, green tinted windows, green hubs, rubbed floor under canopy, decals on sides, yellow and green 'BP', inside green shield, green lettering Exploration, 6 wheels, 3 axles. Made 1967

	£3	£3	£2	£1
Flaw: smooth floor under canopy, one decal missing on one side of above model	£25	£10	£5	£3
4a Smooth floor under canopy, green hubs, decals. Made 1968	£4	£3	£2	£1
4b Ribbed floor, green hubs, sticker decals. Made 1969	£3	£2	£1	75p

No 61 Fifth issue Alvis Stewart White body, bright green hubs, Exploration decal in large green letters, green shield and BP sign in gold, superfast wheels, length 62 mm, 96:1. Made 1971

	£5	£4	£2	£1
5a Flaw: only four wheels, green hubs, back two wheels have dark red hubs	£25	£10	£7	£5

No 61 Sixth issue Blue Shark Purple streamlined body, silver motor in rear, white driver, gaily coloured orange square decal, with number 86 in white, red arrow decal flashes on hood, black

THE GOLDEN YEARS OF LESNEY 61

	Mint boxed	Mint unbxd	Good cond	Fair cond
mounting on sides of motor, superfast wide racing wheels, length 76 mm, 63:1. Made 1972	£2	£1	50p	25p
6a Deep purple body, golden motor, white driver, red arrow decal with number 86 on hood in white with orange background square, superfast wide wheels. Made 1973	£2	£1	50p	25p
6b Deep metallic blue, white driver, silver motor, orange and red flash decals on hood, number 86 in white letters on orange square, superfast wide racing wheels. Made 1974	£2	£1	50p	25p
Flaw: Number 68 in white letters on dark orange square	£25	£10	£5	£3
6c As 6B with slight colour change in motor mounting at sides, in rear of car. Made 1975	£1	50p	15p	10p

No 61 Seventh issue Blue Shark Darker metallic blue, lighter silver plastic motor at rear, orange arrows and number 86 decal on orange square on hood, white driver, very wide superfast racing type wheels. Made 1976

	Mint boxed	Mint unbxd	Good cond	Fair cond
No 62 First issue general service lorry Olive green, silver headlights and front bumper, metal tow hook, three axles, six black plastic wheels, crimpled axles, thick knobby treads, 'TT' scale, 110:1. Made 1959	£10	£5	£2	£1
1a Black plastic wheels, round axles, thick knobby treads	£7	£4	£2	£1
1b Grey plastic wheels, knobby treads. Made 1961	£7	£5	£2	£1
1c Lighter green, grey plastic wheels, length 70 mm, 110:1, TT scale. Made 1962 and 1963	£5	£3	£2	£1
Flaw: 2 black plastic wheels on front and 4 grey plastic wheels on rear, part of bumper missing	£50	£25	£10	£5

No 62 Second issue TV service van Golden yellow, green tinted windows, silver grill, headlights and front bumper, off white plastic rear sliding door, red plastic ladder, antenna and TV sets, red 'Rentaset' decals snap base and TV mast on top of van, length 64 mm, 75:1, OO scale, black plastic wheels, knobby treads

| | £15 | £10 | £3 | £2 |

	Mint boxed	Mint unbxd	Good cond	Fair cond
2a Red Rentaset decals, snap base, black plastic wheels, fine treads	£10	£7	£3	£2
2b Cream body, red access, silver grill, headlights and bumper bar, black plastic base, red Rentaset decals, rivet front base, black plastic wheels, fine treads. Made 1966	£15	£10	£5	£3
2c Deep cream coloured body, bright red decals and accessories, red ladder and TV mast, yellow tinted windows, black plastic wheels, fine treads. Made 1967	£15	£10	£5	£3
2d Cream body, red ladder and access, silver grill, headlights and bumper bar, golden tinted windows, Radio Rental decals, black plastic wheels, knobby treads, length 64 mm, 86:1. Made 1968	£25	£15	£7	£5
Flaw: No decals and no paint on back plastic sliding door, short bumper	£50	£25		

No 62 Third issue Mercury cougar Metallic lime green, red interior and tow hook, metal base and tow slot, opening doors, opening doors, silver plastic hubs, grill and headlights, black plastic tyres, fine treads. Made 1969

	£5	£3	£2	£1
3a Metallic light lime green, black plastic wheels. Made 1970	£5	£3	£2	£1
3b Metallic bright green with yellow tint, silver grill, headlights and bumpers, superfast wheels instead of ordinary black plastic which makes this model a rare find. Last model made at end of 1970	£10	£7	£2	£3

No 62 Fourth issue Mercury cougar Yellow green, green or light green body, red interior and tow hook, silver motor in hood, metal grill and base with Rat Rod sticker decal on sides in red on green background, nicknamed 'Dragster', length 76 mm, 86:1. Made 1971

| | £4 | £3 | £2 | £1 |
| Flaw: motor missing and no decals on doors | £10 | £5 | | |

No 62 Fifth issue Renault 17TL Red, blue tinted windows, off white interior, black grill and base,

62   THE GOLDEN YEARS OF LESNEY

	Mint boxed	Mint unbxd	Good cond	Fair cond
open doors, superfast wheels, length 76 mm, 56:1. Made 1974	£3	£2	£1	50p
5a Metallic rose red, off white interior, blue tinted windows, blue green and orange decals on hood. Made 1975	£1	50p	20p	10p
5b Medium metallic red, number 6 decal in black ringed white circle with green and orange flashes on hood, length 75 mm Made 1976				
No 63 First issue service ambulance Olive green, silver headlights and front bumper, large red cross inside white circle on sides, grey plastic wheels, knobby treads, crimpled axle, TT scale, 100:1. Made 1959	£10	£5	£3	£2
1a Black plastic wheels, crimpled axle, knobby treads. Made 1960/61	£7	£5	£3	£2
1b No silver trim, black plastic wheels, crimpled axle, knobby treads	£7	£5	£3	£2
1c Very dark green, small red cross on sides, grey plastic wheels. Made 1961/62	£7	£5	£3	£2
1d With or without silver trim, black plastic wheels, knobby treads, round axle, small hole under rear axle. Made 1963/4	£10	£7	£5	£3
Flaw: Red cross decal on one side only	£25	£10		
Flaw: Decals missing completely	£25	£10		
1e No silver trim, small hole under rear axle, black plastic wheels, round axle. Made 1964	£7	£5	£3	£2
No 63 Second issue fire fighting crash tender Red, silver grill, bumper, silver metal base and rear, white plastic lettering on side 'Airport Crash Tender', white plastic ladder with hose on roof; 3 axles, 6 black plastic wheels, length 61 mm 93:1, TT/HO scale. Made 1965	£10	£5	£3	£2
2a Silver or gold nozzle on roof, white plastic springs under first and third axle	£7	£5	£3	£2
2b Gold nozzle on roof, no plastic springs, square holes under axles, some models have brace under first axle and square hole under third	£7	£5	£3	£2

	Mint boxed	Mint unbxd	Good cond	Fair cond
2c Gold nozzle on roof, brace under first axle, hole filled in under third axle. Made 1968	£7	£5	£3	£2
No 63 Third issue Dodge crane truck Golden yellow, yellow tinted windows, black grill, bumper and base, moveable crane with hook, 6 black plastic wheels, three axles, length 76 mm, 86:1 Made 1969	£7	£5	£2	£1
3a Yellow body in various shades, green tinted windows, yellow or red plastic hook on moving crane, some models with half base plates and tow hooks, superfast wheels. Made 1970-72	£4	£3	£2	£1
No 63 Fourth issue Freeway gas tanker Rose pink body and chassis, off white tanker, Burmah decals on sides, smaller decal in orange and blue at each side, blue grill, black base, 6 wide superfast wheels, fine treads, length 77 mm, 90:1 Made 1973	£2	£1	50p	25p
4a Orange body, otherwise as above. Made 1974/5	£1	50p	20p	10p
4b Very dark red, clear white tanker, very black Burmah letters on sides, small red, black and blue decal at front end of tanker, length 78 mm Made 1976				
No 64 First issue Scammell breakdown truck Olive green, double cable hook, 6 grey plastic wheels, knobby treads, crimpled axle, TT scale, 98:1 Made 1959	£10	£5	£3	£2
1a Grey plastic wheels, knobby treads, round axle. Made 1961-62	£10	£5	£3	£2
1b Black plastic wheels, crimpled or round axle, knobby or fine treads, green hook, length 67 mm, 98:1 Made 1963-65	£10	£5	£3	£2
Flaw: Missing hook and four grey plastic wheels on rear and two black plastic wheels on front. Made 1965	£25	£10		
No 64 Second issue MG 1100 Green body, off white interior with driver and dog, metal grill, headlights, bumper and base, tow hook, black plastic wheels, length 66 mm, 57:1 Made 1966	£5	£4	£2	£1

THE GOLDEN YEARS OF LESNEY 63

	Mint boxed	Mint unbxd	Good cond	Fair cond
2a Light and medium green, otherwise as above. Made 1967 and 68	£5	£4	£2	£1
Flaw: In 1969 there were quite a number of paint flaws One of the best examples is no paint on body at all and missing driver	£25	£10		
No 64 Third issue MG 1100 Metallic blue, driver, silver grill, headlights, bumpers, hubs and licence plate, white interior, superfast wheels Made 1970	£5	£3	£2	£1
3a Dark metallic blue. Made 1971	£4	£3	£2	£1
No 64 Fourth issue Slingshot dragster Pink body, white driver, silver motor, gaily coloured decal with flower pattern in red, black and yellow on nose, number 9 in yellow circle, large wide superfast racing wheels at back and small racing wheels at front, length 67 mm, 68:1 Made 1972	£2	£1	50p	25p
4b Deep blue metallic body, silver motor and red mountings, number 9 decal with coloured variation on nose Made 1975	£2	£1	50p	25p
No 64 Fifth issue Fire chief Red, white interior, blue tinted windows, blue roof light, silver motor, silver plastic bumpers and lights, silver grill, yellow and red decals on sides with 'Fire' in deep blue letters, length 77 mm, 65:1 Made 1976				
No 65 First issue Jaguar 3.4 litre saloon Metallic blue, silver grill, bumpers and headlights, dark blue tinted windows, large silver plastic wheels, length 76 mm, 72:1, OO scale Made 1959/60	£10	£7	£5	£3
1a Blue metallic body blue grill and headlights, black bumpers, grey plastic wheels, crimpled axle Made 1961	£10	£7	£5	£3
No 64 Third issue MG 1100 Metallic blue; driver; silver grill, headlights, bumpers; silver hubs, licence plate; white interior; superfast wheels; Made 1970	£5	£3	£2	£1
3a Dark metallic blue; Made 1971	£4	£3	£2	£1
No 64 Fourth issue Slingshot dragster Pink body; black, yellow nose; white driver; silver motor; racing				

	Mint boxed	Mint unbxd	Good cond	Fair cond
wheels: flower pattern in red; No 9 in yellow circle; 76 mm; 68:1; Made 1972	£2	£1	50p	25p
4a Deep purple body; brown No 9 in yellow circle; multi coloured decal; Made 1973	£2	£1	50p	25p
4b Deep blue metallic body; silver motor; red mountings; No 9 decal; coloured nose; Made 1975	£2	£1	50p	25p
No 64 Fifth issue Fire chief Red; white interior; blue tinted windows; blue roof light; silver motor; silver plastic bumpers; lights; silver grill; yellow, red decals on sides; 'Fire' in deep blue letters; 77 mm; Made 1976				
No 65 First issue Jaguar 3.4 litre saloon Metallic blue body; large silver plastic wheels; silver grill; bumpers, headlights; dark blue tinted windows; crimpled axle; 72:1; 00 scale Made 1959/60	£10	£7	£5	£3
1a Blue metallic body; grey plastic wheels; crimpled axle; black bumpers; blue grill; headlights; Made 1961	£10	£7	£5	£3
1b Blue; grey plastic wheels; silver bumpers; silver grill; headlights; silver trim; Made 1961	£10	£7	£5	£3
1c Metallic blue; grey plastic wheels; round axle; blue tail lights	£10	£7	£5	£3
Flaw Paint missing from boot, mudguards; badly finished model	£25	£10		
Flaw No 2; grey plastic wheels at front; black plastic wheels at rear	£25	£10		
No 65 Second issue Jaguar 3.8 sedan Metallic red body; silver grill, headlights, front bumper; off-white interior; opening hood; silver motor; tow hook; silver plastic wheels; knobby treads; 72:1; 60 mm; Made 1962	£10	£7	£5	£3
2a Light metallic red; silver plastic wheels; round axle; knobby treads; Made 1963	£5	£4	£2	£1
2b Grey plastic wheels; round axle; fine treads; 60 mm; 72-1; 00 scale	£5	£4	£2	£1
2c Red; black plastic wheels; round axle; fine treads	£5	£4	£3	£2
2d Bright metallic rose pink; green tinted windows; silver bumpers; grill; headlights; silver motor; silver plastic wheels; Made 1964/65	£10	£5	£3	£2

THE GOLDEN YEARS OF LESNEY

	Mint boxed	Mint unbxd	Good cond	Fair cond
2e Light rose pink metallic body; blue bumpers; blue headlights; blue grill; blue motor; dark interior black plastic wheels; Made for International Year 1966; special blue trim on body	£25	£15	£10	£5
Flaw No motor; missing hood; no rear bumper	£50	£25		
No 65 Third issue Combine harvester Red body; yellow plastic reaper paddles; yellow front hubs; black wheels on rear; 76 mm; 106-1; Made 1968/69	£5	£3	£2	£1
3a Black plastic tyres on front; black plastic wheels on rear; hubs varied in colour between 1970 and 1972	£5	£3	£2	£1
No 65 Fourth issue Saab Sonnet Blue body; orange base, interior; opening boot; yellow decal on hood; 72 mm; 53-1; Made 1973	£2	£1	50p	25p
4a Light blue metallic; fine trim; dark, light blue boot covers. Made 1974	£2	£1	50p	25p
4b Dark rich metallic blue; Made 1975	£2	£1	50p	25p
No 65 Fifth issue Saab Sonnet Dark blue body; orange interior, base; lighter boot cover ribbed; Made 1976				
No 66 First issue Citroen DS19 Orange yellow body; silver grill, headlights, bumpers; grey plastic wheels; black base 66 on raised platform; 00 scale; 74:1; Made 1959	£7	£5	£3	£2
1a Yellow body; grey plastic wheels; silver trim	£5	£3	£2	£1
1b Cream body; Made 1960	£5	£3	£2	£1
1c Yellow; grey plastic wheels; crimpled axle; black rear bumper; 66 mm; 74-1; Made 1961/2	£4	£3	£2	£1
1d Yellow; grey plastic wheels; round axle; silver rear bumper	£4	£3	£2	£1
1e Yellow body; silver plastic wheels; knobby treads; silver grill, headlights, bumpers	£5	£4	£2	£1
Flaw Missing grill; no paint on one side; no headlights	£25	£10		
No 66 Second issue Harley Davidson motor cycle & sidecar Metallic blue; 3 silver wire wheels; black tyres; 66 mm; 41-1; 0 scale; Made 1963; treasured item for any collector; one of the rarest in Lesney range	£1,000	£500		
2a Bronze metallic; Made 1964	£100	£50	£10	£5

	Mint boxed	Mint unbxd	Good cond	Fair cond
2b Copper metallic; Made 1965	£100	£50	£10	£5
2c Metallic bronze; 66 mm; 41-1; 0 scale; Made 1966	£250	£100	£25	£10
Flaw Short handle bars owing to overheating model in mould process	£100	£50		
No 66 Third issue Greyhound bus Silver body; white interior; orange tinted windows; 3 axles; 6 black plastic wheels; white background decal; 76 mm; 159-1; Made 1967	£10	£7	£4	£2
3a Clear windows; blue outlines, letters on decal; also has tinted windows	£7	£5	£3	£2
3b Tinted windows; grey outline, letters on sticker decal	£5	£4	£2	£1
No 66 Fourth issue Greyhound bus Metallic silver grey; blue greyhound decals; white interior; golden tinted windows; Made 1969	£10	£5	£3	£2
4a Dog on decal at rear of lettering	£25	£10	£7	£5
4b Silver grey body; 6 black plastic wheels; dog faces front; tow slot; Made 1970	£6	£4	£2	£1
4c Superfast wheels; yellow base; also light yellow base; final issue; Made 1971	£5	£4	£2	£1
Flaw windows missing, no decals on sides, very light metallic colouring; 4 superfast wheels; two non-superfast	£50	£25		
No 66 Fifth issue Mazda RX 500 Orange body, white base, chassis; silver motor; silver interior; boot opening at rear; superfast racing wheels; 72 mm; 59-1; Made 1972	£2	£1	50p	25p
5a Deep orange; purple tinted windows; white base; opening boot; Made 1973/74	£2	£1	50p	25p
5b Dull orange body; dark tinted windows; silver plastic motor; silver base; Made 1975	£1	50p	20p	10p
No 66 Sixth issue Mazda RX 500 Bright red metallic body; white base; rear bumpers; opening boot; gaily coloured decal green, red, white; superfast racing wheels; Made 1976				
No 67 First issue Saladin armoured car Olive green; rotating gun turret; 3 axles; 6 grey plastic wheels; 57 mm; 75-1; 00 scale; knobby treads; Made 1959	£10	£7	£5	£3

THE GOLDEN YEARS OF LESNEY

	Mint boxed	Mint unbxd	Good cond	Fair cond
1a Dark green; medium green, light green colours; grey plastic wheels; knobby treads; crimpled axles; Made 1960/61/62	£7	£5	£2	£1
1b Black plastic wheels; crimpled axles; knobby treads; also round axle; Made 1963/64	£7	£5	£2	£1
No 67 Second issue Saladin armoured car Black plastic wheels; round axles; fine treads; 6 large black plastic wheels; revolving gun turret; 57 mm; 86-1; Made 1965/66/67	£5	£3	£2	£1
No 67 Third issue Volkswagen 1600 TL Light red body; silver metal grill, bumpers, headlights, base, tow slot; off white interior; opening doors; silver trim; 54 mm; 61-1; Made 1968	£6	£3	£2	£1
3a Dark red; silver plastic hubs; black plastic tyres; also medium red colour; Made 1969	£5	£4	£2	£1
3b Red; silver plastic hubs; black plastic tyres; maroon roof rack	£10	£5	£2	£1
No 67 Fourth issue Volkswagen 1600 TL Metallic purple; silver plastic hubs; black plastic tyres; Made 1969	£6	£4	£2	£1
4a Metallic purple; superfast wheels; silver headlights; 3 silver spotlights; silver bumper bars; number plate; silver trim; Made 1970	£5	£4	£2	£1
4b Dark violet; superfast wheels; also medium violet and light violet; Made 1971	£4	£3	£2	£1
4c Metallic hot pink; Made 1972	£4	£3	£2	£1
4d Shocking pink; no colour on headlights; spotlights, or bumpers; missing doors; Made 1973	£25	£10		
No 67 Fifth issue Hot Rocker Light metallic lime; silver motor; silver grill, headlights, bumpers; superfast racing wheels; 76 mm; 59-1; Made 1974	£2	£1	50p	25p
5a Dark olive green; also light mustard; Made 1975	£1	50p	20p	10p
5b Metallic red; silver motor; silver grill, headlights, bumpers; off white interior; 77 mm; 59-1; Made 1976				
No 68 First issue Austin Mk II radio truck Olive green; silver headlights, bumper; grey plastic wheels; crimpled axle; knobby treads; 87:1; HO scale; Made 1959	£10	£7	£5	£3

	Mint boxed	Mint unbxd	Good cond	Fair cond
1a Grey plastic wheels; round axle; knobby treads; Made 1961/62	£7	£5	£3	£2
1b Black plastic wheels; crimpled axle; knobby treads; black base; also round axles, 64 mm; 87-1; HO scale; Made 1963/64/65	£6	£5	£3	£2
1c Black plastic wheels; round axle; fine treads; with or without silver trim; Made 1965	£7	£5	£3	£2
No 68 Second issue Mercedes coach White, orange body; clear windows; white interior; silver grill; headlights; bumper; golden trim; open axles; black plastic wheels; 72 mm; 90-1; HO scale; Made for International Year 1966	£10	£7	£5	£3
2a Green; silver trim; three shades; very rare; Made 1966	£100	£50	£10	£5
2b Orange, deep red orange; open axles	£10	£5	£3	£2
2c Orange red; closed axles; black plastic wheels; also medium and light orange red; golden line trim; Made 1967/68/69	£7	£5	£2	£1
No 68 Third issue Porsche 9.10 Metallic light red; yellow tinted windows; silver metal tail lights, exhaust base; superfast wheels; black 68 on hood; 76 mm; 54-1; Made 1970; also in light and medium, dark red	£5	£4	£2	£1
Flaw in Medium red; two superfast wheels on rear; non superfast on front; no decals	£25	£10		
3a Medium metallic red; superfast wheels; 68 decal; Made 1971	£4	£2	£1	50p
3b Medium metallic red; wide racing wheels; Made 1972/73	£3	£2	50p	25p
3c White; silver V decal; broomstick; various shades; Made 1975	£1	50p	25p	10p
No 68 Fourth issue Cosmobile Deep metallic blue; silver motor, trim; orange yellow base; 73 mm; 61-1; wide superfast wheels; Made 1976				
No 69 First issue Nestle's delivery truck Dark maroon body; silver grill, headlights, bumpers; gold decal; grey plastic wheels; 57 mm; 91:1; HO scale; Made 1959/60	£25	£10	£7	£5

THE GOLDEN YEARS OF LESNEY

	Mint boxed	Mint unbxd	Good cond	Fair cond

1a Dark maroon; dark grey plastic wheels; Made 1960 — £25 £10 £7 £5

1b Light maroon; round axles; light grey plastic wheels — £25 £10 £7 £5

1c Deep red body; full silver trim; golden decals; Made 1963 — £30 £15 £10 £5

Flaw Missing decal on one side; doors solid; windows filled in — £100 £50

1d Light red; light grey plastic wheels; 57 mm; 91-1; HO scale; Made 1964; final model to be made — £30 £25 £10 £7

No 69 Second issue Hatra tractor shovel Orange body; movable shovel; square block under rear axle; orange hubs; grey plastic tyres; 82 mm; 100-1; TT scale; Made 1965 — £10 £5 £3 £1

2a Orange; red hubs; black plastic tyres; Made 1966 — £7 £5 £2 £1

2b Yellow; red hubs; black plastic tyres; hole over rear axle; also yellow hubs; Made 1967 — £5 £3 £2 £1

Flaw Orange body; yellow shovel; black knobby plastic tyres — £50 £10

Flaw Yellow body, orange shovel, black plastic tyres — £50 £10

No 69 Third issue Rolls Royce coupe Silver shadow in rich dark blue; metallic; reddish brown seats; dash; steering wheel; tow hook; orange tinted windshield; trunk opens; silver metal grill, headlights, bumper; superfast wheels; 79 mm; 64-1; Made 1970 — £10 £5 £2 £1

3a Dark blue, medium blue, light blue; light tan rear cover; superfast wheels; black base; silver trim; tan cover; Made 1971/72 — £7 £5 £3 £1

3b Gold body; tan cover; yellow base; also grey base; opening boot; black cover; silver grill, headlights, bumper; Made 1973 — £10 £7 £5 £3

Flaw No paint on body; missing grill — £50 £20

No 69 Fourth issue Turbo Fury Metallic red; white driver; arrow flashes on front No 69; wide superfast wheels; large turbo jets at rear; 76 mm; 63-1; Made 1974/75 — £2 £1 50p 25p

4a Dark metallic red; No 69 decal; 77 mm; Made 1976

No 70 First issue Thames Ford estate car Red, grey; dark grey plastic wheels; silver grill, headlights, bumpers; round axle; knobby treads; no windows; OO scale; 76:1; Made 1959 — £15 £10 £7 £5

1a Dark red, light grey; dark grey plastic wheels; round axles; knobby treads; no windows; Made 1960 — £15 £10 £7 £5

1b Turquoise blue, yellow top; dark grey plastic wheels; round axle; knobby treads; no windows; 55 mm; 76:1; Made 1961 — £7 £5 £3 £2

1c Turquoise, yellow; light grey plastic wheels; green tinted windows; Made 1962 — £5 £4 £2 £1

1d Dark blue, buttercup; silver plastic wheels; knobby treads; clear plastic windows; 54 mm; Made 1963 — £5 £4 £2 £1

1e Dark green, golden yellow; silver plastic wheels; clear windows; silver trim; Made 1964 — £10 £5 £3 £2

1f Light green, golden yellow; black plastic wheels; knobby treads; green tinted windows; Made 1965 all above models made with and without silver rear bumpers — £5 £4 £2 £1

Flaw Windows filled in; no paint on bottom half; slight paint near cab — £50 £25

No 70 Second issue Grit spreader Red cab, chassis; pale yellow body; blue tinted windows; silver metal grill; bumper; pull shoot for unloading; 66 mm; 85-1; HO scale; Made 1966 — £10 £5 £3 £2

2a Black plastic wheels; pull shoot; Made 1967/68/69 — £5 £3 £2 £1

No 70 Third issue Grit spreader Bright yellow body; dark red cab, chassis; silver headlights, grill, base; non superfast wheels; Made 1970 — £5 £3 £2 £1

3a Pale yellow body; rose pink cab, chassis; superfast wheels; silver grill, bumpers; Made 1971 — £4 £3 £2 £1

Flaw Half cab missing; chassis badly warped; no wheels; Made 1971 — £50 £25

No 70 Fourth issue Dodge dragster Bright rose pink body; off white interior; gold tinted windows; silver grill, headlights, bumpers; superfast wheels; snake decals; lift off body; silver motor; 79 mm; 64-1; Made 1972 — £2 £1 50p 25p

THE GOLDEN YEARS OF LESNEY

	Mint boxed	Mint unbxd	Good cond	Fair cond

4a Mauve; wide superfast wheels; silver motor, grill, exhausts; lift off body; Made 1973/74 — £1 · 50p · 25p · 10p

4b Dark pink, medium pink, light pink; silver motor; Made 1975/76

No 71 First issue Austin 200 gal water truck Olive green; silver headlights, front bumper; spare tyre; knobby black plastic wheels; round axle; HO scale; 87:1; Made 1959/60 — £10 · £5 · £3 · £2

1a Square front base; hole in rear does not go right through — £7 · £5 · £2 · £1

1b Square front base; hole goes through tank — £7 · £5 · £2 · £1

1c Hole goes through tank on most models; 66 mm; 87-1; Made 1961/62/63 — £6 · £4 · £2 · £1

No 71 Second issue Jeep pick up truck Medium red; green seats; steering wheel; red grill, headlights, bumper bars; black plastic wheels; round axles; knobby treads; 66 mm; 74-1; 00 scale; Made 1964 — £10 · £7 · £5 · £3

Flaw Missing steering wheel, seats; only three wheels — £50 · £25

No 71 Third issue Jeep pick up truck Red; silver grill; headlights; front bumpers; clear plastic windows; green interior; green plastic door springs; black plastic wheels; round axle; 66 mm; 74-1; 00 scale; Made 1965 — £7 · £5 · £2 · £1

3a White interior; white plastic door springs; Made 1966 — £7 · £5 · £2 · £1

3b White interior; white plastic door springs; patent number; white interior with metal springs; Made 1967/68 — £5 · £3 · £2 · £1

Flaw Missing doors; no windows — £25 · £10

No 71 Fourth issue Ford heavy wreck truck Red cab, crane; white base, chassis; white base; decals; 3 axles; 6 wheels; yellow dome light on roof; silver headlights, grill, bumper bar; black plastic wheels; 76 mm; 78-1; Made 1969 — £7 · £5 · £2 · £1

4a Medium, light, dark red; amber dome light; windows; red plastic hook; black plastic wheels; Made 1970 — £5 · £4 · £2 · £1

4b Amber dome; windows; yellow plastic hook; also green dome; green tinted windows; small sticker decals; red plastic hook; black plastic wheels; Made late 1970 — £5 · £3 · £2 · £1

4c Green dome; windows; dark blue circle decal; superfast wheels; Made 1971 — £5 · £3 · £2 · £1

4d Green dome; windows; light blue circle decal; light red plastic hook; superfast wheels; Made 1972 — £4 · £3 · £2 · £1

Flaw superfast wheels front only; short crane; warped body — £25 · £10

No 71 Fifth issue Jumbo jet Blue; silver forks; engine, exhausts; black plastic wheels; blue seat; red emblem; plastic figure between front forks; 70 mm; 50-1; Made 1973 — £2 · £1 · 50p · 25p

5a Blue base, forks, seats; orange emblem; figure; silver motor, exhausts; small plastic wheels; Made 1974 — £1 · 50p · 20p · 10p

No 71 Sixth issue Jumbo jet Dark royal blue frame; orange figure; silver motor, exhausts; 70 mm; 50-1; Made 1975/76

No 72 First issue Fordson tractor (power major) Blue; tow hook; silver grill; open front axle; small front wheel; large rear wheels; bright red hubs; decals; 62:1; Made 1959 — £7 · £5 · £2 · £1

1a Solid grey plastic wheels on front; rear grey plastic tyres; orange hubs — £6 · £5 · £2 · £1

1b Solid black plastic wheels on front; fine treads; black plastic tyres; orange hubs — £6 · £5 · £2 · £1

1c black plastic tyres; orange hubs — £4 · £3 · £2 · £1

1d Medium blue, bright yellow; black plastic tyres; yellow hubs; 54 mm; 56-1; Made 1967 — £4 · £3 · £2 · £1

No 72 Second issue Standard jeep Deep golden yellow body; black bumper, base; red plastic interior; tow hook; yellow hubs; black plastic tyres; 60 mm; 63-1; Made 1968/69 — £6 · £4 · £2 · £1

2a Yellow, orange body; superfast wheels; balloon tyres; silver headlights; black grill, base; Made 1970/71 — £6 · £4 · £2 · £1

68 THE GOLDEN YEARS OF LESNEY

	Mint boxed	Mint unbxd	Good cond	Fair cond
No 72 Third issue Hovercraft SRN6 White, black; union jack decals; 76 mm; 190-1; Made 1972	£2	£1	50p	25p
3a Royal blue base, white body; lighter union jack decals; red propeller; Made 1973/74/75	£1	75p	50p	25p
3b Dark blue base; grey cheaper plastic version; red propeller; 77 mm; Made 1976				
No 73 First issue 10 ton pressure refueller Airforce blue; airforce decal; 6 grey plastic wheels; silver front bumper; 3 axles; knobby treads; 126:1; Made 1959/60	£7	£5	£3	£2
No 73 Second issue RAF 10 ton pressure refueller RAF blue; 6 silver grey plastic wheels; silver grey plastic bumper; RAF decal; 67 mm; 126:1; Made 1961	£6	£4	£2	£1
2a Light bluish green body; silver bumper; silvery grey plastic wheels; thick round brace; Made 1962	£6	£4	£2	£1
No 73 Third issue Ferrari racing car Red; decals; No 73; silver metal exhaust; steering wheel; wire metal wheels; black plastic tyres; black base; 66 mm; 62-1; Made 1963	£6	£4	£2	£1
3a Bright red; grey driver; also white driver; Made 1964/65	£6	£4	£2	£1
3b Deep maroon body; white driver; blue tinted wire racing wheels; shield decal missing; No 7 on side; rare model	£100	£50		
3c Rose pink body; blue tinted wire wheels; white driver; 67 mm; Made 1966/67/68	£4	£3	£2	£1
No 73 Fourth issue Mercury commuter Rich green; silver headlights, grill, bumper bars; off white interior; two dogs; tow slot; No 55 or 73 on base; silver hubs; black plastic tyres; 79 mm; 72-1; Made 1969	£6	£4	£2	£1
4a Metallic lime green; black plastic tyres	£4	£3	£2	£1
4b Light metallic lime green; superfast wheels; Mercury separated from base; Made 1970	£4	£3	£2	£1
4c Light metallic lime green; superfast wheels Mercury connected to base	£4	£3	£2	£1
Flaw inside completely missing; no paint on roof, or hood	£25	£10		
No 73 Fifth issue Mercury commuter Bright metallic red; white interior; wide superfast racing wheels; silver grill, bumpers, headlights; orange decal; Made 1972	£7	£5	£3	£2
No 73 Sixth issue Mercury commuter station wagon Dull yellow green; plastic bumpers; headlights; grill; off white interior; superfast wheels; no decals; Made 1973	£1	75p	50p	25p
No 73 Seventh issue Weasel Lime green; black revolving plastic gun, turret; wide superfast wheels; black base; 72 mm; 52-1; Made 1974	£2	£1	50p	25p
7a Dark green; jet black turret, base; Made 1975	£1	50p	20p	10p
7b Bright metallic green; jet black turret; base; wide wheels; silver trim; 72 mm; 52-1; Made 1976				
No 74 First issue Mobile canteen White; cream interior; tow bar; opening counter door; open axle; grey plastic wheels; decals on roof; grey trim line around centre of body; 00 scale; 80:1; Made 1959	£200	£50	£10	£8
1a Silver grey; blue interior, base; tow bar; opening counter; open axle; decals; Made 1960	£25	£15	£7	£5
1b White; grey plastic wheels; spread rivet; tow bar; also comes in cream; Made 1961	£15	£10	£7	£5
1c Metallic pink; blue interior; tow bar; decals; 66 mm; 80-1; rare model; Made 1961	£50	£25	£10	£5
1d Silver grey; blue interior; silver plastic wheels; knobby treads; 67 mm; Made 1964/65	£10	£7	£3	£2
1e Silver body; grey plastic wheels; fine treads; dark blue interior; rivet base; tow bar; Made 1966	£5	£4	£3	£2
Flaw No decals; no windows	£50	£10		
Flaw Odd wheels; one grey solid tow bar; missing interior	£100	£25		
Flaw No interior; blue part of interior and other part white; no hinged door; solid piece across centre of counter; no decals	£250	£50		
No 74 Second issue Daimler bus Cream; white interior; black plastic wheels; snap in front base; rivet at top of first rear window; decals; 76 mm; 121-1; Made 1976	£50	£25	£5	£3

THE GOLDEN YEARS OF LESNEY 69

	Mint boxed	Mint unbxd	Good cond	Fair cond
Flaw decals dropping down at angle; large letters; small decal missing; no paint on part of roof;	£100	£50		
Flaw Decals missing completely	£100	£50		
2a Cream; decals; Made 1968	£50	£25	£5	£3
2b Cream; black plastic wheels; decals; 76 mm; Made 1968	£50	£25	£5	£3
No 74 Third issue Daimler bus Green; black plastic wheels; decals; Made 1969	£50	£25	£5	£3
No 74 Fourth issue Daimler bus Red; black plastic wheels; non superfast wheels; Made late 1969	£25	£10	£7	£5
4a Red, medium, dark, light; superfast wheels; decals; Made 1970	£10	£5	£3	£2
4b Red fluorescent body; red base; superfast wheels; varying shades; Made 1971/72	£7	£5	£3	£2
Special Issue Cyprus Airways decal	£500	£200		
Flaw No wheels; no wheelbase or base plate	£25	£10		
No 74 Fifth issue Toe Joe Lime green; bright green crane; red plastic tow hook; twin dome lights; orange tinted windows; wide superfast wheels; 76 mm; 66-1; Made 1973	£1	75p	50p	25p
Light yellow body; lime green crane; red plastic hook; wide superfast wheels; Made 1974	£1	75p	50p	25p
Dark lime green metallic body; bright green crane; red hook; Made 1975	£1	50p	20p	10p
Flaw No crane; tow hook; dome lights	£5	£2		
No 74 Sixth issue Toe Joe Lime green; twin amber flash lights on roof; orange tinted windows; dark green tow bar; orange hooks; wide super wheels; 74 mm; Made 1976				
No 75 First issue Ford Thunderbird Cream, orange; silver grill, headlights; red tail lights; metal base; tow hook; green tinted windows; open axles; 00 scale; 80:1; Made 1959	£10	£7	£5	£3
1a Cream, pink; silver trim; Made 1960	£10	£5	£2	£1
1b Dark blue; silver plastic wheels; knobby treads; white top; dark oak side panels; Made 1961/62	£10	£5	£3	£2

	Mint boxed	Mint unbxd	Good cond	Fair cond
1c Dark blue, pink; black base; silver plastic wheels; grey plastic wheels; Made 1963	£5	£3	£2	£1
1d Cream, pink or orange; black base; grey plastic wheels; knobby treads; 67 mm; 80-1; 00 scale; Made 1964	£5	£3	£2	£1
No 75 Second issue Ferrari Berinetta Light blue; white interior; tow hook; clear windows; silver grill, headlights; metal bumper, base, tow slot; rare model; 72 mm; 62-1; Made 1965	£25	£10	£5	£3
2a Light green; also medium green; wire wheels; black plastic tyres; Made 1966	£6	£4	£2	£1
2b Dark green; wire wheels; black plastic tyres; also comes with silver hubs; Made 1967/68	£6	£4	£2	£1
2c Light green; black plastic wheels, tyres; flaw with wrong wheels; Made 1969	£25	£10		
No 75 Third issue Ferrari Berinetta Red; silver grill, headlights, bumpers; stick shift between seats; superfast wheels; 72 mm; 62-1; Made 1970	£5	£3	£2	£1
3a Without stick shift between seats; some models with red grill; Made 1970	£4	£3	£2	£1
No 75 Fourth issue Alfa Carabo Deep pink body; black louver rear; off white interior; yellow base; wide superfast racing wheels; 76 mm; 56-1; Made 1971	£2	£1	50p	25p
4a Light pink body; wide wheels; black louver rear; off white interior; orange or yellow base; Made 1972	£2	£1	50p	25p
4b Deep purple body; blue-black louver rear; off white inside; white base; Made 1973/74	£1	75p	30p	15p
4c Metallic pink; three shades; blue-black louver rear; yellow orange base; off white interior; Made 1975	£1	50p	25p	10p
4d Flaw No paint on body; louver rear uneven	£25	£10		
No 75 Fifth issue Alfa Carabo Shocking pink; black, lemon or orange triangles on louver at rear; decals on drivers side; Colours can be found to be reversed; wide superfast racing wheels; yellow orange base; 76 mm; 56-1; Made 1976				

MODELS OF YESTERYEAR

	Mint boxed	Mint unbxd	Good cond	Fair cond
The royal state coach, processional with a team of 8 horses gold with 4 riders in red jackets with black caps, red trimming and white trimming, white horses with grey manes, the figure of the Queen inside. With all traces and measuring approximately 381 mm long. Made 1947/48	£250	£200	£100	£75
State processional coach with a team of 8 horses In silver with 4 riders in red jackets and black caps, white horses with grey manes and red and white trimming. Both of the above models are very rare. Made 1948/49	£300	£200	£100	£75
The miniature coronation coach with four riders in red and black, horses with red and white trim, approximately 127 mm long, original price 2/11d. Made 1953	£75	£55	£35	£25
King size milk float orange with orange driver and brown horse, approx. 102 mm long. Made 1947/49	£50	£25	£15	£10
King size bulldozer in yellow and orange with driver, approx. 102 mm long. Made 1947/51	£50	£25	£15	£10
King size prime mover with tractor orange and yellow with driver, approx. 102 mm long. Made 1947/51	£50	£25	£15	£10
Y1 Allchin 7N H P. traction engine first issue green, gold trim, 6 spokes on black left side fly wheel, 10 spokes on small front wheels and 16 spokes on the larger rear wheels, 00 scale, 80:1, very rare model. Made 1956	£50	£30	£15	£10
1a red front and rear wheels with crimpled axles, otherwise details as above	£50	£30	£12	£10
1b unpainted treads on rear wheels and red on front wheels, crimped axle	£50	£30	£12	£10

	Mint boxed	Mint unbxd	Good cond	Fair cond
1c red front wheels with round axles and red rear wheels	£50	£30	£15	£10
1d red front wheels with unpainted rear treads and round axle	£50	£30	£15	£10
1e silver boiler door and silver trim, bright red front wheels, round axles, unpainted rear treads	£50	£30	£15	£10
1f silver trim, very smooth rear treads, a very rare model indeed. Only a few were manufactured in England	£250	£100	£15	£10
Y1 Second issue 1911 model T Ford commonly known as the 'Tin Lizzie' and one of the greatest sellers in the history of the motor car. Red body with red wheels, black roof and gold steering wheel, black seats and radiator grill, gold radiator frame, headlights, handbrake, horn and wire wheels, black plastic tyres, length 76 mm, 42-1 0 scale. Made 1965	£25	£15	£7	£5
2a red body, black roof with duel control handles. Golden trim, two holes in the base, very rare model. Made in England	£50	£30	£15	£10
2b gold trim, single control handle, two holes in base	£25	£20	£7	£5
2c red and black with gold trim, no holes in base	£25	£22	£10	£5
2d with black plastic steering wheel, no holes in base	£25	£22	£10	£5
2e silver body, with silver metal steering wheel, black roof, red seats and radiator grill	£50	£30	£15	£10
2f silver body, silver metal steering wheel, red roof, seats and radiator grill	£50	£30	£15	£10
2g silver body, black plastic steering wheel and red or black roof	£50	£35	£12	£10

The Royal State processional coach and the miniature Coronation coach

Models of yesteryear

Y1 third issue model T Ford white body with red plastic roof and red seats, red mudguards and gold trim, black superfast tyres with silver spokes. Made 1973 — £5 £2 £1

Y2 first issue B type bus in deep red with metal driver, open rear stairway to tan open deck seats, silver grill, Dewars decal on upper side, General decal on lower sides, Oakeys decal at the rear and decals at upper front and rear for various destinations, TT scale, 100:1. Made 1956 — £50 £30 £18 £15

1a blue driver with deep red colour of bus, 4 small windows above and 4 large lower windows on each side, blue metal driver, crimpled axle. Very rare and made in England — £100 £50 £25 £15

1b dark red or light red, blue metal driver, metal wheels, 8 small windows above and 4 large lower windows on each side, crimpled axle — £50 £30 £20 £15

1c and 1d blue driver, and one with black driver, metal wheels with round axle and black metal wheels with round axle — £50 £30 £10 £7

Y2 second issue 1911 Renault metallic green, red seats, gold dash board, windshield and side lights, headlights, licence plate, handbrake and spoked wheels with a 4 prong tyre mount with black plastic tyres, length 76 mm, 40-1 0 scale. Made 1964 — £15 £10 £4 £2

2a unpainted dash board, windshield frame, lights and tyre mount, metal steering wheel with two holes in base — £15 £10 £3 £2

2b unpainted tyre mounts — £7 £5 £3 £2

2c all gold trim — £7 £5 £3 £2

2d 3 pronged tyre mount — £7 £5 £3 £2

2e black plastic steering wheel — £7 £5 £3 £2

2f silver body with four or three pronged tyre mounts — £25 £15 £10 £5

Please note that there are as many as ten various shades for the body colour alone; the wheels of these models are mostly silver or gold with the occasional colours being mixed, and horn buttons are various sizes

Y2 Third issue 1914 Prince Henry Vauxhall red with silver grey hood, red radiator with white seats, gold headlights, radiator, frame, windshield, left battery box, rear gas tank, black plastic steering wheel, spoked wheels, black plastic tyres, length 89 mm, 47-1. Made 1971 — £10 £7 £5 £3

Flaw silver spoked wheels on front with only a slight touch of gold, and silver on radiator badly finished. — £25 £10

Y2 Fourth issue 1914 Prince Henry Vauxhall metallic blue body with silver hood, gold dash, windshield, headlights, radiator frame, black radiator and black steering wheel, silver plastic wheels with superfast tyres, length 89 mm, 47-1. Made 1973. The shade of blue may vary in this model. — £1.05

Y3 first issue 1907 London E class tram car red body, gold headlights, front and rear stairs lead to enclosed deck, upper side decals 'News of the World', lower decals 'London Transport', end side decals 'Buy Lesney Toys', front and rear decals 'City', length 130:1. Made 1956 — £40 £30 £7 £5

72 LESNEY — MODELS OF YESTERYEAR

	Mint boxed	Mint unbxd	Good cond	Fair cond
1a details as above but with 'Dewars' decal, very rare as only a few were made	£100	£50		
1b cream roof, grey bumpers, opening under stair wells, metal wheels, 'News of the World' decal	£40	£30	£10	£5
1c cream roof, grey bumpers, closed under stair wells, metal wheels	£40	£30	£7	£5
1d white roof, grey bumpers, metal wheels	£40	£30	£7	£5
1e cream roof, grey bumpers, black plastic wheels	£40	£30	£5	£3
1f white roof, closed opening leading to upper deck, black plastic wheels	£40	£30	£5	£3
1g white roof, white bumpers	£40	£30	£5	£3
Flaw 1h roof pink with almost no red on body, also no decals, very special	£500	£100		

Please note that the above Y3 models, have decals in various shades and different thicknesses of lettering. Proofs are detachable and can be switched from one model to another. Some of the base plate lettering reads either front to back or back to front. There are also various shades of red on the bodywork.

	Mint boxed	Mint unbxd	Good cond	Fair cond
Y3 second issue 1910 Benz limousine cream body with light green roof, gold dash, windshield, hand break, three pronged tyre mount, radiator shell and headlights, also golden spoked wheels, black plastic tyres, length 83 mm, 54-1 O scale, the original price was 5s 0d. Made 1966	£25	£15	£7	£5
2a rich golden cream colour, dark blue radiator, dark olive green seats, hood and roof, otherwise all gold trim as above. Made 1967	£25	£15	£7	£5
2b rich cream colour with green roof, red seats and radiator. This model had a dark golden orange roof and radiator. Made 1968	£50	£25	£10	£5
Y3 third issue 1910 Benz limousine light green body, yellowy green roof, gold dash and trim, red seats and interior, fine gold trim on body and black radiator, ribbed tyres and golden spoked wheels, original price 5s 6d. Made 1969	£15	£10	£7	£5

1907 London E Class tram car

	Mint boxed	Mint unbxd	Good cond	Fair cond
3a green body, black roof, red seats, all gold trim as above, red radiator, black steering wheel, price went up to 5s 11d. Made 1971	£10	£7	£5	£3
3b metallic blue body, light green roof, red seats and radiator, also with black roof. Made 1972	£10	£7	£5	£3
Y3 fourth issue 1934 Riley MPH metallic red, white seats, silver plastic radiator, headlights and silver horns, silver windscreen, black steering wheel, silver spoked wheels, length 99 mm, 35-1. Made 1973	£2	£1		
4a light pink body, white plastic seats, headlights, radiator, horns, licence plate, silver plastic wheels. Made 1974, 1975, 1976	£1			
Y4 first issue Sentinel steam wagon blue body, black chassis, gold trim left side tank and top of stack, decal inside red border line in orange writing 'Sand and Gravel Supplies', decal on nose of cab, metal wheels, crimpled axle, TT scale, 100:1. Made 1956	£75	£50	£10	£5
1a orange lettering with lighter blue background and maroon border, metal wheels, crimpled axle. Made 1957/58/59	£75	£50	£7	£5
1b black blue tinted body, metal wheels with crimpled axle and almost white lettering, rare. Made 1960	£100	£75	£10	£5

LESNEY — MODELS OF YESTERYEAR

	Mint boxed	Mint unbxd	Good cond	Fair cond
1c yellow lettering, bright red border, larger rear metal wheels with round axle	£75	£50	£7	£5
1d knobby treads and orange or yellow lettered decal, black plastic wheels with round axle, last model made	£50	£40	£7	£5
Y4 second issue Shand Mason horse drawn fire engine red body, gold boilers, three black figures with gold helmets, white horses with grey manes and tails, black spoked wheels, 'Kent Fire Brigade' decals in yellow and gold on horses collars, length 89 mm, 63-1. Made 1961	£50	£20	£10	£5
2a white horses with 'London Fire Brigade' decals in yellow with black border and yellow 72. Made 1962/63	£50	£20	£10	£5
2b black horses with white manes and tails, black helmets on figures, 'London Fire Brigade' in yellow with black border and 72. Made 1964	£50	£40	£10	£5
2c black horses, gold helmets on figures, 'London Fire Brigade' with silver boilers	£50	£40	£10	£5
2d black horses, gold helmets and boilers, gold outlined border on decals	£75	£50	£10	£5
2e black horses, gold boilers, black helmets on figures with half rounded brace under boilers	£50	£40	£7	£5
Flaw: missing boiler and no brace under boilers	£150	£75		
Y4 third issue 1909 Opel coupe white body, tan roof, white steering wheel, gold dash, radiator shell, lights, handbrake and spoked wheels, black plastic tyres, length 79 mm, 38-1, original price 5s 0d. Made 1967	£25	£20	£7	£5
3a white body, maroon seats, red radiator with smooth fire wall, black steering wheel, roof attached by metal pins. Made 1968	£10	£7	£5	£3
3b dark cream body, red seats, red radiator. Made 1969	£10	£7	£5	£3
3c red seats and radiator, roof supports attached to seat pins. Made 1970	£10	£7	£5	£3
3d white body, no holes in base, red seats, red radiator, golden spoked wheels with golden trim. Made 1971	£15	£10	£7	£5

	Mint boxed	Mint unbxd	Good cond	Fair cond
3e all silver trim except red radiator and seats. Made 1972	£10	£7	£5	£3
Y4 fourth issue 1909 Opel coupe orange body, white radiator and gold metal trim, silver plastic wheels, superfast tyres, length 79 mm, 38-1. Made 1973	£5	£3	£2	£1
Y4 fifth issue 1930 Duesenberg model J metallic red body, black hood, silver radiator, headlights, bumpers and horns, black interior, silver spoked wheels, superfast tyres, length 114 mm, 43-1. Made 1976				
Y5 first issue 1929 Le Mans Bentley green, red painted seats, silver steering wheel, silver blower, metal solid wheels and black rubber tyres, No. 5 decal with crimpled axles, O scale, 55:1. Made 1957/58	£50	£45	£10	£5
1a silver radiator shell with grey rear cover, all green seats, crimpled axle, very rare. Made 1959	£50	£45	£10	£5
1b gold radiator shell with green rear cover, crimpled axle. Made 1960	£50	£45	£10	£5
1c gold radiator shell with green rear cover, round axle. Made 1961	£50	£45	£10	£5
Y5 second issue 1929 S Bentley green, silver grill, headlights, windscreen, wire wheels, black plastic tyres, steering wheel and union flag decal on doors, No 5 decal in white circle on sides, length 89 mm, 52-1 O scale. Made 1963	£30	£25	£7	£5
2a metallic green, also dark green, green seats and tonneau. Made 1964	£30	£25	£5	£3
2b light green body, light green seats and tonneau. Made 1965	£30	£25	£5	£3
2c metallic green, red seats, tonneau, No. 3 decal on sides and two holes in base. Made 1966	£30	£25	£10	£5
2d red seats and tonneau, No. 5 decal with two holes in base. Made 1967	£30	£25	£5	£3
2e red seats and tonneau, No. 6 decal on sides, two holes in base	£50	£40	£15	£10
2f red seats and tonneau, decalled No. 5, no holes in base. Made 1968	£30	£25	£5	£3
2g silver body, red seats and tonneau, No. 5 decal	£25	£15	£10	£5

74 LESNEY – MODELS OF YESTERYEAR

	Mint boxed	Mint unbxd	Good cond	Fair cond
Y5 Third issue 1907 Peugeot Metallic blue body; black steering wheel; green gold wheels; black tyres; red seats; dark blue hood; golden windscreen, dashboard; gold trim; 89 mm; 43-1; 7/2d; Made 1969	£15	£10	£5	£3
Y5 fourth issue 1907 Peugeot Yellow body, black roof and red seats, red radiator, black steering wheels on most of these models, gold trim, orange side windows, black plastic tyres, spoked wheels. The body of some of the models had a rich dark golden tan. Made 1970/71	£10	£7	£5	£3
Flaw: no paint on body and no spare tyre or prongs to hold same	£100	£50		
Y5 fifth issue 1907 Peugeot copper bronze body, black seats, copper bronze hood, gold trim and black base, silver spoked wheels, superfast tyres. Made 1973	£5	£3	£2	£1
5a orange pink body, 89 mm 43-1. Made 1974/75/76	£1			
Y6 First issue AEC Y type lorry Light grey body; dark grey driver; steering wheel; silver radiator, hood handles; decals; metal wheels; crimpled axle; TT scale 100:1; Made 1957/58	£50	£40	£7	£5
1a dark grey with metal wheels and crimpled axle	£50	£35	£10	£5
1b dark grey, grey hood handles and bright red outline, metal wheels, round axle	£50	£40	£7	£5
1c rich dark grey, black plastic wheels, very rare as only a few were made for a special consignment	£100	£50		

	Mint boxed	Mint unbxd	Good cond	Fair cond
Y6 second issue 1923 type 35 Bugatti blue body, red dash and seats, gold radiator, steering wheel, gas cap, hand brake, spoke wheels, black plastic tyres, length 79 mm, 46-1 O scale. Made 1962	£50	£30	£10	£5
2a very light sky blue, black seats, black base, radiator, hand brake, steering wheel and petrol cap, deep gold spoked wheels, deep red No. 6 decal in white circle. Made 1963	£50	£30	£10	£5
2b blue, fine treads with unplated or gold wheels, black plastic tyres, No. 6 decal. Made 1964	£50	£30	£10	£5
2c royal blue, white dash with blue seats, black plastic tyres, decalled No. 9 a genuine flaw	£250	£100		
2d red body, black seats, gold radiator with petrol cap, golden spoked wheels, red No. 6 in white decal. Made 1966	£30	£25	£10	£5
2e light pink body, red cap, black seats, red radiator, gold wheels, No. 6 decal in black in pinkish white circle, very rare in this class. Made 1967	£100	£25		
2f red, white dash, red seats with no trim, black plastic tyres, decalled No. 6. Made 1967	£20	£15	£7	£5
Y6 third issue 1913 Cadillac deep blue body, fine gold trim lines, dark brown roof, orange seats and radiator, black steering wheel, golden windscreen, headlamps, side lamps, rear spare wheel holder, golden spoked wheels, length 87 mm, 48-1, original price 5s 6d. Made 1968	£25	£10	£7	£5
3a metallic gold body with no holes in base, otherwise details as above. Made 1969	£10	£7	£5	£3
3b metallic gold, holes in base and maroon roof. Made 1970	£10	£7	£5	£3
Flaw: metallic gold and the letter 1 is missing from the 1913 on base	£25	£15	£10	£5
Y6 fourth issue 1913 Cadillac metallic green body, black hood, dark orange seats, gold radiator, headlights, side lights, windscreen, spare wheel at the rear, 4 silver plastic spoked wheels, black superfast tyres, length 89 mm, 48-1. Made 1973	£2	£1		

1929 Le Mans Bentley

AEC Y type lorry

LESNEY — MODELS OF YESTERYEAR

	Mint boxed	Mint unbxd	Good cond	Fair cond
4a very dark green body, blue cover, orange radiator, orange seats, gold trim, white silver plastic spoked wheels. Made 1974	£2	£1		
Flaw: missing petrol tank from running board	£25	£10		
4b very light green body, black roof, golden radiator, golden spoked wheels, this model should be looked after as the normal wheels are silver	£100	£50		
4c dull green body with pale orange seats and interior. Made 1976				
Y7 first issue 4 ton Leyland lorry dark brownish maroon, rear window openings and side window openings, silver grill, metal wheels, crimpled axle, cream roof, 'W & R Jacob & Co Ltd' decal in yellow, outlined in orange, two lower lines of lettering 'by Royal Appointment to His Majesty the King'. 'Biscuit Manufacturers' length 100:1, TT scale. Made 1957/58	£50	£40	£15	£10
Flaw: 'By Royal Appointment to His Majesty the King' missing decal	£250	£100		
1a dark maroon, silver radiator and grill, very light cream roof, metal wheels and crimpled axle, full decals. Made 1959	£50	£40	£15	£10
1b light brown, silver grill, metal wheels and crimpled axle, decals completely missing from one side	£500	£100		
1c light brown, silver grill, metal wheels and crimpled axle with full decals. Made 1960	£50	£40	£10	£5
1d light brown, silver grill, metal wheels with round axle, light fine lettering on decals	£50	£40	£10	£5
1e dark brown with golden sheen, white roof, silver grill, metal wheels, dark gold lettering on decals, length 70 mm. Made 1961	£50	£40	£15	£10
Flaw: missing 'Jacob' and also 'Ltd' and 'Biscuit Manufacturers' decal on one side	£200	£50		
Flaw: Box with part of van missing; bad fault of printers when boxes were being made; and although there are several others flaws with regard to the boxes themselves; this is one of the rare finds.	£100	£25		

	Mint boxed	Mint unbxd	Good cond	Fair cond
Y7 second issue 1913 Mercer raceabout sports car lilac, very faint in colour, black seats, gold radiator, shell, headlights, steering wheels, handbrake, petrol cap, double spare tyres, silver spoked wheels, black plastic tyres, length 82 mm, 48-1 O scale. Made 1962	£50	£40	£15	£10
2a deep lilac, running board runs in at an angle until it meets the body in a point, model has no holes in the base, lilac grill, headlights, radiator top, steering wheel, handbrake, golden spoked wheels, black plastic tyres with knobby treads. Made 1963	£50	£40	£15	£10
2b deep purple body, also metallic light purple effect on some models, silver or gold spokes, black plastic tyres, fine treads. Made 1964/65	£50	£40	£15	£10
Y7 third issue 1913 Mercer raceabout sports car golden yellow body, black seats, fully golden trim, gold radiator, golden steering wheel, dark steering column, double spare tyres on rear, spoked wheels, black plastic tyres with knobby treads, length 82 mm 48-1 O scale. Made 1966 for the International Year	£50	£40	£10	£5
3a deep golden orange, deep gold radiator, black seats, spoked wheels. Made 1967	£40	£30	£10	£5
3b copper bronze body, rare model as only a few were made for a special market, the golden trim on this model was of a far superior quality than many of the other Yesteryear models	£100	£75	£25	£10
Flaw: very short steering wheel, almost on dash board, missing headlights and part of grill, single tyre on rear	£250	£100		
Y7 fourth issue 1912 Rolls Royce golden yellow body with black roof, very smooth, red seats and radiator, black steering wheel, golden headlights and other accessories, black plastic tyres, golden rims and spoked wheels, spare wheel on driver's side, length 96 mm, 48-1, original price 6s 6d. Made 1968	£50	£30	£25	£15
4a silver, red running board and mudguards, red roof, gold trim on wind shield frame, headlights, radiator shell, horn, tool box and extinguisher, smooth roof or smooth roof with part red grid or full roof with all grid on top, golden spoked wheels,				

76 LESNEY — MODELS OF YESTERYEAR

	Mint boxed	Mint unbxd	Good cond	Fair cond
spare wheel on drivers side with gold frame. Made 1969/70	£20	£15	£10	£5
4b model as above with grey or medium grey roof	£15	£10	£7	£5
4c grey or red ribbed roof	£20	£15	£10	£5
4d red ribbed roof with maroon seats or red ribbed roof with red seats and no holes in the base	£20	£15	£10	£5
4e silver body with grey ribbed roof	£50	£25	£15	£10
4f all gold body with red ribbed roof	£50	£25	£15	£10

Y7 fifth issue 1912 Rolls Royce dark fawn or orange brown metallic body, gold frame, headlights, horn, side lights, black seats, pink partly ribbed and partly smooth roof, pink running board, mudguard with spare wheel, silver spoked wheels, superfast tyres, length 96 mm, 48-1. Made 1973 — £1

Y8 first issue Morris Cowley 1926 'Bullnose' light and medium brown body, golden radiator, frame and grill, spare tyre mounted on right side, crimpled axle with knobby treads, copper spoked wheels, OO scale, 50-1. Made 1958 — £75 £50 £15 £10

1a tan body, missing radiator, copper wheels, spare wheel badly made, almost solid spokes — £100 £50

1b tan body, silver wheels, round axle and knobby treads. Made 1959 — £75 £50 £15 £10

1c light tan body, silver wheels, round axle and knobby treads. Made 1961 — £75 £50 £15 £10

1d orange tanned body, gold radiator, golden spoked wheels, length 64 mm, 50-1 O scale. Made 1962 — £75 £50 £15 £10

Y8 second issue 1914 Sunbeam motor cycle with Milford sidecar silver body with green seats in sidecar, black seat on cycle, silver spoked wheels, length 67 mm, 34-1. Made 1963/64/65 — £15 £10 £7 £5

2a black seats inside sidecar and black seat on motor cycle, black plastic tyres. Made 1966/67/68 — £15 £10 £7 £5

Y8 third issue 1914 Sunbeam with Milford sidecar black motor cycle with black sidecar, fine golden trim on motor cycle, silver motor, black spoked wheels, length 67 mm, 34-1, very rare, originally 5s 4d. Made 1969 — £100 £75 £50 £25

Flaw: very short handlebars and black motor cycle with deep green seats in sidecar — £250 £100

All black 1914 special Sunbeam motor cycle with black sidecar — £1,000 £500

Y8 fourth issue 1914 Stutz metallic red body, green seats and radiator, tan roof, black plastic steering wheel, gold windowframe, headlights, radiator shell, horn, handbrake, tool box tops, also golden plated horn and a very fine gold line trim on red body, also medium red body, luggage and spare tyre, spoked wheels, black plastic tyres, length 86 mm, 48-1, original price 7s 0d. Made 1970 — £15 £10 £7 £5

4a gold coloured gas tank in rear. Made 1971 — £15 £12 £10 £7

4b black seats and bronze gas tank in rear. Made 1972 — £15 £10 £7 £5

Flaw: silver spoked wheels in front and gold spoked wheels in rear — £50 £25

Flaw: radiator frame and petrol tank frame with most of the other gold trim almost completely covered with metallic red — £50 £25

Y8 fifth issue 1914 Stutz metallic blue body, black roof, white seats, gold window frame, headlights and radiator shell, horn, handbrake and tool box tops, plastic steering wheel, silver spoked wheels, length 86 mm, 48-1. Made 1973 — £1

Flaw: metallic blue paint on tool box tops, spoked wheels almost filled in, bad factory fault, patchy black roof — £15 £5

Y9 first issue Fowler showmans engine deep dark maroon, cream roof with copper roof supports and boiler door, small front wheels and large rear wheels,

LESNEY — MODELS OF YESTERYEAR

	Mint boxed	Mint unbxd	Good cond	Fair cond
orange spokes with unpainted smooth treads, Lesneys Modern Amusements decal at top sides, star decal and square box decal on lower sides, OO scale, 80-1. Made 1958	£50	£40	£25	£10
1a maroon with cream roof, copper roof supports and gold boiler. Made 1959	£40	£25	£15	£10
1b maroon with cream roof, gold roof supports and gold boiler. Made 1960	£40	£25	£15	£10
1c maroon with cream roof, gold roof supports and header, gold boiler. Made 1961	£30	£25	£15	£10
1d red body with cream roof, gold supports and header, gold boiler. Made 1962	£50	£40	£15	£10
1e red body with white roof, gold supports and header, gold boiler. Made 1963	£75	£50	£15	£10
1f red body with white roof, gold supports and header, gold boiler and black base. Made 1964	£75	£50	£15	£10
1g red body with white roof, gold supports and header, silver boiler and red base. Made 1965	£75	£50	£15	£10
1h red with white roof, gold front support and header, silver rear support and header and silver boiler. Made 1965	£75	£50	£25	£15
1j red with white roof, silver front support and header, gold rear support and header with gold boiler, also has silver supports with header and silver boiler. Made for the 1966 International Year	£60	£45	£20	£10
Flaw: no paint on roof and gold supports very short, no gold paint on header and boiler	£100	£50		
Y9 second issue 1912 Simplex Light green or medium green body, tan roof, red seats, red radiator, gold trim, spoked wheels and spare tyre mounted on three prongs, black plastic tyres and deep gold spoked wheels, length 96 mm, 48-1, original price 6s 0d. Made 1968	£15	£10	£7	£5
2a dark green body with roof supports attached to metal pins. Made 1969	£12	£10	£7	£5
2b dark green body, dark tan roof with supports attached to red seat pins	£12	£10	£7	£5

	Mint boxed	Mint unbxd	Good cond	Fair cond
2c dark green body with a rich tan darkened hood with a nice textured pattern	£15	£12	£9	£7
Flaw: Silver spoked wheels instead of gold	£50	£20		
Y9 third issue 1912 Simplex Rich golden body colour, black roof, black seats, full golden trim, dark red grill, spoked wheels, black thick ribbed tyres. Made 1970	£25	£20	£15	£10
3a gold body, black textured roof, rich dark red seats and radiator grill. Made 1971/72	£10	£9	£7	£5
Y9 fourth issue 1912 Simplex Light red metallic body, orange seats, black hood, gold and silver trim on headlights, radiator frame, orange radiator grill, spare wheel prongs, silver spoked wheels, black plastic superfast tyres, length 96 mm, 48-1. Made 1973	£1			
Y10 first issue 1908 Grand Prix Mercedes Off white, green painted seats, silver chain drive, silver exhaust pipes, silver steering wheel, gold trim, spoked wheels with black plastic knobby tyres and two spare tyres on the rear, O scale, 54-1. Made 1958	£75	£50	£20	£15
1a green seats, knobby treads with crimpled axle	£75	£50	£15	£10
1b Knobby treads with round axle, green seats. Made 1959/60	£30	£25	£12	£10
1c cream body, black seats, golden chain drive and pipes, silver tint in the golden spoked wheels. Made 1961	£100	£50		
1d white body, dull in parts with silver drive chain and black seats, golden trim and silver pipes, double spare tyres, length 76 mm, 54-1, O scale. Made 1962	£100	£50		
Y10 second issue 1928 Mercedes Benz 36/220 White with red plastic seats, tan dashboard and floor, silver trim on horn, headlights, grill and exhausts, spare tyre on rear and spoked wheels with golden trim, length 96 mm, 52-1, O scale, very special and rare. Made 1964	£100	£50	£15	£10

LESNEY — MODELS OF YESTERYEAR

	Mint boxed	Mint unbxd	Good cond	Fair cond
2a silver spoked wheels, silver trim. Made 1965	£15	£12	£7	£5
2b brilliant white with silver sheen, bright red seats and all silver trim, two spare wheels on the rear.	£25	£20	£15	£10
2c one spare tyre with two holes in base. Made 1967	£25	£20	£7	£5
2d one spare tyre with no holes in the base. Made 1968	£15	£10	£7	£5
2e silver sheen with two spare tyres and almost white body covered with silver sheen, golden spoked wheels	£25	£20	£15	£10
2f golden body with part silver trim and deep dark red seats	£100	£50		

Y10 third issue Rolls Royce Silver Ghost metallic lime green with metallic brown chassis and base, black plastic steering wheel, golden trim, spoked wheels, black plastic tyres, length 92 mm, 51-1, round axles. Made 1970

	Mint boxed	Mint unbxd	Good cond	Fair cond
	£18	£15	£10	£7
3a maroon seats, brace at end of front springs on baseplate. Made 1971	£15	£12	£10	£5
3b Dark red seats and model without braces. Made 1971	£15	£12	£10	£7
3c Metallic gold body with golden spoked wheels and golden trim, metallic gold base and mudguards, deep pink seats. Made 1972	£12	£10	£6	£4

Y10 fourth issue 1906 Rolls Royce Silver Ghost White metallic body, metallic pink chassis, gold trim, black plastic seats and spare wheel, silver spoked wheels with black plastic superfast tyres, length 92 mm, 51-1. Made 1973

£1

Y11 first issue Aveling & Porter Steam roller Green body, black roof, small metal rollers, red spokes with gold trim, large rear wheels, OO scale 80-1. Made 1958

	£50	£40	£15	£10
1a dark green body, black roof, black supports, black flywheel. Made 1959	£50	£40	£15	£10

	Mint boxed	Mint unbxd	Good cond	Fair cond
1b green roof with black supports, brownish flywheel. Made 1960	£50	£40	£10	£7
1c green roof with green roof supports and black flywheel, grey metal rollers and large grey rear wheels. Made 1961	£50	£40	£10	£7
1d light blue green tinted body, almost orange spokes and very dark red inside rollers, blue black fly wheel, orange spokes in large wheels with deep red trim, length 86 mm, 80-1, OO scale, very rare. Made 1962	£80	£60	£20	£15
1e green roof, black flywheel, small front rollers, silver spokes and large rear wheels with red spokes, length 79 mm, 80-1, OO scale. Made 1963	£80	£45		

Y11 second issue 1912 Packard Landaulet red with black seats and hood, two holes in base, metal steering wheel, black spare tyre with four prongs on mount, black plastic tyres with golden spoked wheels, shield decal on doors, length 82 mm, 50-1, O scale. Made 1964

	£15	£10	£7	£5
2a silver fire wall, floor and front grill with gold trim. Made 1965	£15	£10	£7	£5
2b silver spare tyre mount and gold trim, also with silver floor and silver headlights, silver grill, silver steering wheel	£50	£25	£15	£10
2c silver steering wheel with gold trim	£15	£10	£7	£5
2d full silver trim on all parts, silver wheels, exceptionally rare	£100	£50		
2e deep red body, all gold trim	£25	£15	£10	£5
2f gold trim with silver three pronged tyre mount	£20	£10	£7	£5
Flaw: No shield on door and missing part of steering wheel	£25	£10		

Please note that this model comes in many wheel combinations of silver and gold

Y11 third issue 1938 Lagonda Drophead coupe metallic pink body and hood, metallic bronze chassis, mudguards and running board, black hood

LESNEY – MODELS OF YESTERYEAR

at back with black radiator grill, silver headlights, gold windscreen and radiator grill surround, silver plastic wheels, superfast tyres, length 112 mm, 43-1. Made 1973

Y12 first issue London horse drawn bus red body, tan upper open deck seats with tan driver, black hat, two brown horses with white manes and tails, fine gold collars, metal spoked wheels, decals on upper deck 'Liptons Tea', length 83 mm, 100-1, TT scale. Made 1959

	Mint boxed	Mint unbxd	Good cond	Fair cond
	£50	£40	£15	£10
1a dark brown horses, one rivet holding tow bar, dark yellow lettering on lower decal. Made 1958/59	£50	£40	£15	£10
1b brown horses, bright yellow lettering on decals. Made 1960	£50	£40	£10	£7
1c brown horses with two rivets holding tow bar. Made 1961	£50	£40	£10	£7
1d almost black horses with very dark tan seats and driver, no paint on wheels, prominent white letter decal on upper deck with blue and white decals next to driver, dark gold lettering decals on lower. Made 1964	£100	£50		

Y12 second issue 1909 Thomas Flyabout metallic dark blue body, tan roof, maroon seats and maroon radiator, gold headlights and sidelights, black plastic steering wheel with gold column, gold trimmed spoked wheels and spare tyre mount on three prongs, black plastic tyres, length 102 mm, 48-1, original price 5s 6d. Made 1968

	£15	£12	£9	£7
2a dark blue body, brown seats, roof supports are attached to metal pins on the side with two holes in the base. Made 1969	£15	£10	£7	£5
2b very light or medium blue with very dark red seats and two holes in base. Made 1970	£15	£10	£7	£5
2c dark blue body, red seats, roof supports attached to seat pins with no holes in base, square platform on base with 1909 Thomas Flyabout. Made 1971	£25	£15	£10	£7
2d light blue body, medium or dark blue with a black textured or dark tan textured roof	£20	£15	£10	£7

London horse drawn bus

American 4-4-0 locomotive

Flaw: two golden spoked wheels on the rear and two silver spoked wheels on the front. Made 1972   £50   £25

Y12 third issue 1909 Thomas Flyabout bright metallic red body with black hood, white plastic seats, gold windscreen, headlights and side lights, black steering wheel and black radiator grill, gold radiator frame, gold pronged spare tyre holder, silver spoked wheels, black plastic superfast tyres, length 102 mm, 48-1. Made 1973

Y13 first issue American 4-4-0 locomotive dark green body with maroon front section and chassis, golden trim on three stacks, red boiler door, golden face of lamp, black metal wheels, TT scale, 112-1. Made 1959

	£50	£40	£10	£7
1a silver headlight with gold trim on three stacks, deeper maroon on chassis and front section, metal wheels, orange yellow decals	£50	£40	£10	£7
1b maroon headlight with gold trim on three stacks, yellow decal	£50	£40	£15	£10
1c maroon headlights with gold front stack only yellow decal	£50	£40	£7	£5
Flaw: very short stacks on two smaller sizes	£75	£45		

The above engine has rather a fantastic history, when it first came into public service. Apart from being boxed and herded into the crude passenger carriages each occupant between the age of 7 years and 70, had to make a mark or sign the back of the ticket before they were allowed to ride. They had to promise to help in all ways that would be required by the conductor of

	Mint boxed	Mint unbxd	Good cond	Fair cond

the train. Even to the extent of fighting off bandits, Indians and wild animals, gather fuel for the train and help in any way possible so that the train would reach its destination in comparative safety.

Y13 second issue 1911 Daimler Yellow body, black fenders, black base and steering wheel, golden trim and black radiator, black seats, spare tyre on right side, gold spoked wheels and black plastic tyres, length 83 mm, 45-1, O scale. Made 1966 — £15 £10 £7 £5

2a black seats, wheel well and 5 section steering wheel, gas pedal — £15 £10 £7 £5

2b red seats, wheel well, 5 section steering wheel, gas pedal — £12 £9 £7 £5

2c red seats, open wheel well, 5 section steering wheel, gas pedal — £12 £10 £7 £5

2d red seats with no hole in base, depth greater in orange body — £10 £8 £6 £5

Flaw: missing spokes in front wheels, solid wheels — £100 £50

Y13 third issue RAF Crossley Tender blue body, orange tan plastic cover on cab and body, golden trim, red radiator grill, spare wheel, silver plastic spoked wheels with black plastic tyres, red cross decals on cover and RAF decals on sides, length 98mm, 47-1. Made 1973 — £5 £3 £2 £1

3a very deep blue body with dark red cross — £2 £1

Flaw: only two silver spoked wheels on front, gold on rear — £25 £10

Y14 first issue The Duke of Connaught (GWR) dark green with dark brown front section and chassis, black base, gold boiler and small stack with gold trim, six small metal wheels and two large black metal wheels, decal on the large wheel housing 'Duke of Connaught', length 76 mm, 130-1. Made 1959 — £50 £40 £10 £7

1a dark golden colour on boiler door, gold stacks and gold tool boxes, golden grease boxes on lower sides, 4 on each side, green metal wheels, that normally should all be black. Flaw — £250 £100

	Mint boxed	Mint unbxd	Good cond	Fair cond

1b as 1a with black metal normal wheels — £50 £40 £10 £7

1c light green body, gold boiler door with gold stacks and no gold on grease or tool boxes at sides — £50 £40

1d gold boiler door, gold stacks, golden grease boxes and brace under rear wheels. Made 1962 — £50 £40 £7 £5

1e golden boiler door, golden middle stack and no other trim, brace under rear wheels. Made 1963 — £50 £40 £7 £5

Flaw: 1f no gold on boiler door and small wheels missing, heat in moulding melted axle holes and joints — £250 £100

1g very dark green body, deep orange red background, where decal on wheel housing is in golden yellow letters. Made 1964/65 — £50 £40 £7 £5

Y14 second issue Maxwell Roadster 1911 blue green body with black roof, golden trim, bronze gas tank, no holes in base, radiator, metal steering wheel, gold spoked wheels, black plastic tyres, length 82 mm, 49-1, O Scale. Made 1966 — £20 £15 £7 £5

2a two holes in base with gold trim. Made 1967 — £15 £10 £7 £5

2b two holes in base with silver radiator grill, headlights and dash frame. Made 1968 — £15 £10 £7 £5

2c no holes in base with silver radiator grill, dash frame and headlights, silver spoked wheels, rare — £50 £25

2d no holes in base and all parts golden trim. Made 1969 — £15 £9 £7 £4

2e no holes in base, gold trim with red radiator. Made 1970 — £10 £7 £5 £3

2f silver body with red seats, roof and radiator — £50 £25

2g gold body with maroon seats, roof and radiator. Made 1971/72 — £15 £10 £7 £5

2h blue green body, black hood and red seats, gold trim, copper petrol tank, golden spoked wheels, black plastic tyres, original price 7s 0d. Made 1972 — £10 £9 £7 £5

Y14 third issue 1931 Stutz Bearcat metallic lime green body with golden luggage rack at rear, gold windscreen, headlights, horns and bumpers, light

LESNEY — MODELS OF YESTERYEAR 81

	Mint boxed	Mint unbxd	Good cond	Fair cond
red seats, licence plate, black or gold steering wheel and stick, silver spoked wheels, black plastic superfast tyres, length 118 mm, 43½-1. Made 1973	£1			
Y15 first issue Rolls Royce Silver Ghost Metallic green body, black seats, gold radiator shell with green radiator grill, Golden headlights, side lights, and inside windscreen, outside windscreen green metallic, spare tyre on right side of base, golden wheel holder, rich deep golden spoked wheels, length 83 mm, 55-1, O scale. Made 1960	£25	£20	£15	£10
1a silver rear license plate, red tail lights, small opening at rear of running board and side panels, grey plastic tyres, silver wheels. Made 1961	£20	£15	£7	£5
1b green license plate, tail light, small opening, rear of running board and panels, grey plastic tyres, silver wheels. Made 1962	£20	£10	£7	£5
1c black plastic tyres, silver wheels, no opening at rear. Made 1963-64	£15	£10	£7	£5
1d all gold trim, black plastic tyres, gold wheels	£15	£10	£7	£5
1e no holes in base, black plastic tyres, gold wheels. Made 1966	£15	£10	£7	£5
1f rear hole in base covered, black plastic tyres, gold wheels and trim. Made 1968	£15	£10	£7	£5
Y15 second issue: 1930 Packard Victoria metallic bronze, dark brown running boards and chassis, rich maroon running board and rear trunk, red radiator grill, two spare tyres mounted in wheel wells, golden spoked wheels, black plastic tyres, length 108 mm, 46-1, original price 8/6d. Made 1970	£25	£20	£10	£7
2a dark brown seats	£25	£20	£10	£7
2b dark red seats, almost black radiator grill	£25	£20	£10	£7
2c black seats, black running board, black grill and rear trunk	£50	£25		
Y15 third issue 1930 Packard Victoria metallic golden yellow, red radiator, red seats, deep blue hood and luggage box, golden trim with silver spoked wheels and two spare wheels in side walls on running board, superfast tyres, length 108 mm, 46-1. Made 1973	£5	£3	£2	£1

	Mint boxed	Mint unbxd	Good cond	Fair cond
3a metallic lemon, black radiator, black seats and black luggage box, gold trim and silver spoked wheels, grey superfast tyres. Made 1974/75	£25	£15		
3b metallic orange, red seats, red grill, silver spoked wheels. Made 1976	£1			
Y16 first issue Merryweather horse drawn fire engine red body with gold boilers, 3 black figures with gold helmets, white horses with gold collars, white tails and white mane, black spoked wheels, small metal spoked wheels at front and large black metal spoked wheels at rear, London Fire Brigade decals in yellow with black border, stamped Y16 on base and also Y16 correctly marked on the box is all important with regard to this model (a very limited edition), length 89 mm, 63-1. Made 1960	£5,000	£1,000		
Y16 second issue 1904 Spyker veteran automobile pale yellow body, green seats, golden handbrake, metal steering wheel, golden spoked wheels, black plastic tyres with knobby treads, length 83 mm, 45-1, O scale. Made 1961	£25	£20	£15	£10
2a pale yellow, gold radiator shell, horn, side lights and hand brake, fine gold silver tinted spoked wheels, knobby treads. Made 1962	£25	£20	£15	£10
2b pale yellow, gold radiator shell, face of horn, side lights and hand brake, fine treads. Made 1963	£15	£10	£7	£5
2c pale yellow, gold face of horn, side lights and hand brake, with full running board, fine treads. Made 1964	£15	£10	£7	£5
Y16 third issue 1904 Spyker veteran automobile dark yellow body, gold face of horn, side lights and hand brake, dark green seats, deep gold spoked wheels, fine treads. Made 1965	£25	£15	£10	£7
3a dark yellow, green seats, golden horn, gold face of side lights and hand brake. Made 1966	£15	£10	£7	£5
3b silver body, green seats with short side panels on running boards	£20	£15	£10	£7
3c silver body, green seats and long side panels	£20	£15	£10	£7
3d all gold body and full golden trim	£25	£20	£15	£10

LESNEY — KINGSIZE SPECIALS

Kingsize specials

	Mint boxed	Mint unbxd	Good cond	Fair cond
3e yellow body, very light green seats, full golden trim, deep gold metallic spoked wheels, fine treads. Made 1970/71	£15	£10	£7	£5
Y16 fourth issue 1928 Mercedes S S metallic silver body with red running board, mudguard and chassis, golden radiator mount, golden headlights and exhaust and windscreen, black rear box and black radiator with 2 spare wheels in side wells on ribbed running board, golden spoked wheels with black hood, length 108 mm, 45-1, original price 50p. Made 1972	£25	£15	£10	£7
Y16 fifth issue 1928 Mercedes S S metallic lime green body and chassis, black hood, black radiator and black box at rear, fully gold trimmed, silver spoked wheels with superfast tyres, length 108 mm, 45-1. Made 1973	£1			
Y17 first issue 1938 Hispano Suiza all silver metallic body, red seats and red radiator, silver headlights, silver bumpers and black hood and mudguards, silver spoked wheels. Made 1973	£1,000	£100		
1a red metallic body, black mudguards and black hood, silver radiator surround and silver bumpers, headlights and silver spoked wheels with superfast tyres, length 115 mm. Made 1976	£1			

1938 Hispano Suiza

	Mint boxed	Mint unbxd	Good cond	Fair cond
K1 first issue Weatherhill hydraulic shovel yellow body, silver grill, black metal hubs, small decal in front in black background with larger decal in rear with red green background and black lettering, length 57:1, O scale. Made 1960	£4	£3	£2	£1
1a black hubs, grey tyres, length 93 mm, 57-1. Made 1961	£4	£3	£2	£1
K1 second issue 8 wheel tipper truck orange body and cab, silver grill and headlights and front bumper, 4 axles and 8 wheels, thick black decals Hoveringham on sides, length 108 mm, 69-1, OO scale, original price 5s11d. Made 1964	£5	£4	£2	£1
2a deep red cab and body. Made 1965	£5	£4	£2	£1
2b special bronze red cab and orange body, deep bronze red hubs, thicker set dealers decal on sides, Hoveringham, on deeper orange background, made for the International Year 1966	£10	£5	£3	£2
K1 third issue O & K excavator red body with orange tinted windows, silver grey crane and shovel, black sleeves on hydraulic lifters with white interior and red background, white lettered sticker decals, length approx 104 mm, 62-1. Made 1970	£3	£2	£1	50p
3a red hubs with black plastic tyres. Made 1971	£3	£2	£1	50p
3b superfast king wheels	£3	£2	£1	50p
K1 fourth issue O & K excavator red with silver grey lighter plastic crane and shovel, thick super-wheels, decals, 124 mm, 62-1. Made 1976				
K2 first issue Muir Hill Dumper dark red, silver grill, black metal seat and steering wheel, small				

LESNEY — KINGSIZE SPECIALS 83

	Mint boxed	Mint unbxd	Good cond	Fair cond
front wheels and larger rear wheels, small decal above grill, decal on left side of hood, showing vents, 'Muir Hill' decals on sides of tipper, length 76 mm, 56-1, O scale. Made 1960				
1a black metal hubs and grey plastic tyres	£4	£3	£2	£1
1b green metal hubs and black plastic tyres	£3	£2	£1	50p
K2 second issue K-W dump truck yellow body, green tinted windows, silver metal engine and horns, silver hydraulic lifters, black sleeves on lifters, red hubs, 3 axles and 10 wheels, black plastic tyres, length 143 mm, 96-1, TT scale. Made 1965	£5	£4	£3	£2
K2 third issue Scammell heavy wreck truck white body, red dome lights, silver metal bumper and roof horns, silver muffler on sides, red grill and crane, metal hooks with off white plastic cables, white sticker decals, 'Esso', inside blue circle, length 120 mm, 69-1, made in finest die cast metal. Made 1969	£5	£4	£3	£2
3a white body, green tinted windows and orange dome lights, red hubs, other details as above, made of finest diecast	£5	£4	£3	£2
3b gold body with orange tinted windows and dome lights, blue silver tinted hubs, super king wheels. Made 1971	£4	£3	£2	£1
K2 fourth issue Scammell heavy wreck truck gold body, silver horns and orange dome lights, red crane, grill and bumper, grey plastic hook, red steps, super king wheels, white Esso decals. Made 1976				
K3 first issue Caterpillar bulldozer yellow body with engine, tow hook in rear, green rubber treads and moveable blade, Cat D9" decal at top of blade, length 91 mm, OO scale, 70-1	£5	£4	£3	£2
1a metal plastic rollers	£5	£4	£3	£2
1b red plastic rollers	£5	£4	£3	£2
1c red plastic rollers, no blade **flaw**	£25	£10		
K3 second issue Hatra tractor shovel orange with green tinted windows, movable shovel with black sleeves on hydraulic lifters, black spare tyre mounted behind cab on body, white background, black lettered				

	Mint boxed	Mint unbxd	Good cond	Fair cond
decals 'Hatra' on sides, length 150 mm, 61-1. Made 1965	£4	£3	£2	£1
2a red hubs and decals	£4	£3	£2	£1
2b red hubs, black plastic tyres, sticker decals	£3	£2	£1	50p
Flaw: edge of shovel smaller at one end and smooth, no claws	£25	£10		
K3 third issue Massey-Ferguson tractor & trailer red tractor with grey metal interior, tow hook, engine and base, white plastic grill, blue tinted windows, red trailer with yellow tow bar and chassis, black hydraulic sleeves and white plastic tow loop with yellow hubs, length 204 mm, 45-1, original price 10s 6d. Made 1970	£7	£5	£3	£2
K3 fourth issue Mod tractor and trailer yellow seat with silver motor and yellow radiator, blue base, large plastic wheels on rear and small plastic wheels on front with red plastic hubs, superfast wheels, red, white and blue stars and stripes decals on top of tractor and on the sides of yellow and blue trailer, length 198 mm, 45-1. Made 1974				
K4 first issue international tractor red body with red steering wheel and tow hook, red headlights, black plastic tyres with green hubs, white lettered decals on sides, 'McCormick International', b-750 in white circle, length 74 mm, 37-1. Made 1960	£6	£3	£2	£1
1a green metal hubs and tractor with silver headlights	£6	£3	£2	£1
1b red metal hubs and thick tow hook	£6	£3	£2	£1
1c red plastic hubs and thick tow hook	£5	£3	£2	£1
1d red plastic hubs, thick tow hook with hole in base	£5	£3	£2	£1
1e orange plastic hubs, thick tow hook with hole in base	£5	£3	£2	£1
K4 second issue GMC tractor with hopper train dark red cab with blue tinted silver headlights and bumper guards, blue tinted windows with spare tyre mounted behind cab, red hubs, 2 axles, with 6 wheels on tractor attached to first hopper, first hopper has 1 axle with 4 wheels, attached to second hopper by pin and plastic				

LESNEY — KINGSIZE SPECIALS

	Mint boxed	Mint unbxd	Good cond	Fair cond
tow bar, with 2 axles and 8 wheels, red and white sticker decals on hoppers 'Fruehauf', length 286 mm, 67-1. Made 1967	£8	£6	£4	£2
2a Lighter red cab, silver headlights, green tinted windows, green hubs, black plastic tyres,	£25	£15	£10	£5
Flaw: Green cap, green hubs, missing spare wheels, place for spare wheels very short out of line with body of cap	£100	£50		
K4 Third issue Leyland tipper Red cab, chassis, silver plastic grill, quarter base, tinted windows, white interior, silver grey tipper, black plastic sleeves on hydraulic lifters, decal, 8/6d, 115 mm; 56-1, Made 1970	£4	£3	£2	£1
Flaw: Missing petrol tank, very short based mudguards on rear	£50	£25		
3a Red plastic hubs, black plastic tyres, decals	£4	£3	£2	£1
3b Pea green tipper, red orange cab, black plastic tyres, super king wheels	£4	£3	£2	£1
K4 Fourth issue Big tipper Red cab, mustard tipper, super king wheels, white silver mounting on hood of wagon, plastic grill, red front bumper, 117 mm, 69-1, Made 1974	£2	£1	75p	50p
K4 Fifth issue Big tipper Red plastic body, chassis, silver mounting on cab bonnet, plastic headlights, super king wheels, yellow tipper, decals, yellow petrol tank, 118 mm, Made 1976				
K5 First issue Foden tipper truck Orange yellow body, silver grill, decals, metal hubs, black plastic tyres, 108 mm, 74-1, 00 scale. Made 1961	£5	£4	£3	£2
1a Red plastic hubs, black plastic tyres	£4	£3	£2	£1
K 5 Second issue Racing car transporter Green, clear windows, silver grill, headlights, silver grey metal bumper, rear door, base, decals, red plastic hubs; large black plastic tyres, back opens with extended runways to load cars, creviced roof on van, 7/6d, 127 mm, 54-1; Made 1967	£10	£5	£3	£2
2a Light lime green	£9	£7	£5	£4

	Mint boxed	Mint unbxd	Good cond	Fair cond
2b Green hubs, dark green shade on roof of van, rare model, 8/5d, Made 1969	£25	£15	£10	£7
K5 Third issue Muirhill tractor & trailer Yellow, large plastic wheels, 8 axles, 8 wheels, 4 on tractor, 4 on dumper, decals, silver horns, silver radiator grill, motor, grey plastic driver, red, white striped blade which is moveable, red hubs, 242 mm, 50-1, Made 1974/76				
K 6 First issue Allis-Chalmers earth scraper Orange body, red engine, two wire springs on scraper, silver headlights, bumper, decal, 147 mm, 64-1, 00 scale; Made 1960/61	£6	£4	£3	£2
1a Metal hubs, black plastic tyres, Made 1962	£6	£4	£3	£2
1b Red plastic hubs, black plastic tyres	£5	£4	£2	£1
K 6 Second issue Mercedes Benz ambulance White super body, silver metal grill, headlights, bumpers, off white interior, blue tinted windows, blue dome lights, red cross on hood, shield on doors, true guide steering, suspension, opening doors, one patient, silver hubs, 7/6d; 105 mm; 45-1; Made 1968	£10	£7	£5	£3
2a Plastic hubs, black plastic tyres, decals, Made 1969	£7	£5	£3	£2
2b Silver plastic hubs, black plastic tyres, rich cream body, very rare model	£15	£10	£7	£5
Flaw: Missing decals, rear doors fastened with over moulding, part of paint on roof missing, no dome light	£100	£50		
2c off white body, silver plastic hubs, sticker decals, black plastic tyres, Made 1969	£6	£4	£2	£1
K6 Third issue Cement mixer Blue green cab, red radiator grill, red chassis, yellow barrel, red, white stripes, or orange stripes, super king wheels, 146 mm, 58-1, Made 1971	£4	£3	£2	£1
Light blue cab, bright red radiator grill, Made 1974	£4	£3	£2	£1
K6 Fourth issue Motor cycle transporter Blue body yellow roof, base, decals, six wheels super, opening at rear to hold motor cycles, 120 mm, 35-1, Made 1976				
K7 First issue Curtiss Wright rear dumper Yellow body, red motor, 2 axles, four wheels on tractor, one axle,				

LESNEY — KINGSIZE SPECIALS

	Ming boxed	Mint unbxd	Good cond	Fair cond
two wheels on dumper, decals, 150 mm, 00 scale; Made 1960	£5	£4	£3	£2
1a Metal hubs, black plastic tyres	£5	£4	£3	£2
Flaw: Bracket that joins dumper from tractor very short, almost touching ground, also part missing off dumper body at the top	£25	£10		
K7 Second issue S D refuse truck Red cab, chassis, silver grill, headlights, clear plastic windows, silver grey metal tipper, off white interior of cab, decals, 117 mm; 66-1; 7/11d; Made 1967	£5	£3	£2	£1
2a Red hubs, black plastic tyres; decal, No 7 on raised platform on base of model, Made 1968	£4	£3	£2	£1
2b red hubs, clear decal, No 7 on raised platform, tow slot, Made 1969	£4	£3	£2	£1
2c Red hubs, black plastic tyres, decals, tow slot, super king wheels, Made 1971	£4	£3	£2	£1
K7 Third issue Racing car transporter Deep yellow tan body, clear orange tinted body cover, tan chassis, silver horns, silver grill; silver hubs, super king wheels, drop down body at rear, 6 wheels; black plastic tyres, decals, 166 mm, 55-1, Made 1974	£2	£1	50p	
K7 Fourth issue Racing car transporter Bright yellow body, chassis, three axles, six super king wheels, silver horns, bumper, grill, opening doors at sides and rear, decals, 156 mm, 55-1; Made 1976				
K8 First issue Prime mover and transporter with caterpillar crawler Orange cab, silver grill, headlights, green tinted windows, silver grey base, tow hook, decals, orange trailer, black tow bar, orange decals, yellow caterpillar tractor, red motor, green rubber tracks, 12 wheels, 313 mm, 70-1; 00 scale, Made 1962	£15	£10	£7	£5
Flaw: A completely missing from 'Laing' decals on trailer	£50	£25		
1a Very rare model has 'Laing' decals only, Made 1962	£25	£15	£10	
1b Metal hubs on tractor, trailer, metal rollers on caterpillar with K3 on base, Made 1963	£10	£7	£5	£3
1c Metal hubs on tractor, trailer, red plastic rollers on caterpillar with K8 on base	£10	£7	£5	£3

	Mint boxed	Mint unbxd	Good cond	Fair cond
1d Red plastic hubs on tractor, trailer, red plastic rollers on caterpillar, Made 1964	£10	£7	£5	£3
1e Red plastic hubs on tractor, trailer, orange rollers on caterpillar, Made 1966	£10	£7	£5	£3
1f Letters 'Laing' in purple on deep orange body of mover and transporter, black plastic hubs on trailer only, deep maroon engine and rollers on caterpillar, blue letters on white background on doors of prime mover, Made 1966	£25	£15	£10	£5
K8 Second issue Car transporter Turquoise cab, orange trailer, orange hubs, grey plastic tyres, green tinted windows, silver grill, headlights, bumper, decals, 8 large wheels, 8/11d, 209 mm; 73-1, Made 1967	£12	£10	£7	£5
2a Turquoise cab, orange trailer, orange hubs, black plastic tyres	£10	£7	£5	£4
2b Turquoise cab, red plastic hubs	£10	£7	£5	£4
K8 Third issue Car transporter Yellow cab, yellow body, silver trim, bumpers, decals, rare model	£1,000		£200	
Yellow cab, body of transporter, normal decals, Made 1968/69	£10	£8	£6	£4
K8 Fourth issue Caterpillar traxcavator Yellow body, golden yellow bucket, arms, blue driver, white hook, decals, green rubber tracks, black hydraulic bars, rare model, 108 mm, 50-1, Made 1970	£25	£20	£10	£5
4a Yellow body, orange bucket, arms; blue driver, hook, decals, Made 1971	£10	£7	£5	£4
4b Yellow, orange rollers, Made 1972	£10	£7	£5	£4
4c Black rollers	£10	£7	£5	£3
K8 Fifth issue Caterpillar traxcavator Bright yellow body, bright orange bucket, arms, black rubber tracks, black driver, yellow head, 106 mm, Made 1976				
K9 First issue Diesel road roller Green body, red metal rollers, plastic driver, no decals on sides, 96 mm, 57:1, 0 scale, Made 1962	£10	£8	£4	£2
1a Grey plastic driver, decals in two lines, Made 1963	£6	£5	£3	£2
1b Red plastic driver, decals in two lines, Made 1964	£7	£5	£3	£2

LESNEY – KINGSIZE SPECIALS

	Mint boxed	Mint unbxd	Good cond	Fair cond
1c Red plastic driver, decal in one line, Made 1965	£7	£5	£3	£1
K9 Second issue Combine harvester Lime green body, yellow steering wheel, red metal harvester blades, red decals on sides, red plastic hubs, black plastic tyres, green ladder, 140 mm, 80-1; Made 1967	£15	£10	£7	£5
2a Black plastic hubs, shade of red sprinkled, Flaw	£25	£15	£10	£7
2b Red plastic hubs, black plastic tyres, white driver	£10	£7	£5	£3
2c Red plastic hubs, black plastic tyres, white driver, sticker decals, Made 1969	£7	£5	£3	£2
2d Orange body, yellow driver, yellow metal harvester blades, sticker decals, yellow orange hubs, Made 1970	£10	£9	£6	£5
2e Red body, yellow metal harvester blades, tan driver, orange hubs, 55p; Made 1972	£7	£5	£3	£2
K9 Third issue Fire tender Red engine, silver grey plastic ladder, silver horn, trim, super king wheels, Fire Dept decals, clear plastic windows, off white interior, 166 mm, 55-1, Made 1973	£4	£3	£2	£1
3a Bright pink body, grey, black ladder, Made 1976				
K10 First issue Aveling Barford tractor shovel Blue gem stone colour, with greenish tint, red metal seat, steering wheel, double stacks, movable shovel, black plastic tyres, metal hubs, 105 mm; 62-1, Made 1963	£15	£10	£7	£5
Silver blue grey body, silver grey tinted hubs, very rare, Made 1964	£50	£25		
1b Light mint blue, hubs to match, Made 1965	£10	£9	£6	£4
Flaw: Missing stacks, part of steering wheel missing	£100	£50		
K10 Second issue Pipe truck with six interlocking pipes Golden orange body, chassis, silver metal bumper, metal base, exhaust and pipe supports, tow hook, black grill, off white tow loop, 3 axles, 10 wheels on tractor cab, 2 axles, 8 wheels on trailer, dark grey plastic pipes, black, yellow decals on doors, red hubs, thick black plastic tyres, plastic load grips, metal horns, 8/11d, 203 mm, 69-1; Made 1967	£15	£10	£7	£5
2a Rich golden yellow body, Made 1968/69	£12	£9	£6	£5

	Mint boxed	Mint unbxd	Good cond	Fair cond
K10 Third issue Pipe truck Pink body, chassis, 8 wheels on truck, silver bumper, horns, headlights, pipe stays, silver hubs; orange pipes, silver exhaust, black radiator grill, no decals, 11/-d; Made 1971	£9	£7	£5	£3
3a Lemon plastic pipes, dark orange circle decal on brown triangle on door, Made 1972/73	£7	£5	£3	£2
3b Deep purple cab, chassis, 6 yellow pipes, Made 1974	£7	£5	£3	£2
K10 Fourth issue Car transporter Red cab, red transporter, silver grey car loader, black, silver hydraulic stays, gold engine, silver horns, bumpers, hubs, black plastic wheels, 269 mm, 64-1; Made 1976				
K11 First issue Fordson tractor & farm trailer Light silver blue tractor, black plastic wheels, orange hubs, red trailer, orange tint, orange plastic hubs, silver headlights, 158 mm, 42-1; Made 1963	£12	£10	£7	£5
1a Dark blue tractor, silver grill, headlights, red decal, blue chassis on trailer, grey tipper, plastic loop on tow bar, Made 1964	£10	£7	£5	£3
1b Orange metal wheels on tractor, silver steering wheel, orange plastic wheels on tipper, grey sleeves on hydraulic lifters	£9	£7	£5	£3
1c Orange plastic wheels on tractor, silver steering wheels, orange wheels on trailer, grey sleeves on lifters	£7	£5	£3	£2
1d Orange plastic wheels on tractor, blue steering wheel, red plastic wheels on trailer, black sleeves on lifters	£7	£5	£3	£2

4-ton Leyland Jacobs Biscuits lorry

LESNEY — KINGSIZE SPECIALS

	Mint boxed	Mint unbxd	Good cond	Fair cond
1e Red plastic wheels on trailer, square tow hook, red plastic wheels on tractor, black sleeves on lifters	£7	£5	£3	£2
K11 Second issue DAF car transporter Blue cab, grey seats, steering wheel, silver headlights, bumper, four black plastic wheels, red plastic hubs, yellow trailer, black hydraulic lifts, red grips on roof, DAF decals on sides, 12/6d, 229 mm; 66-1, Made 1970	£10	£9	£7	£5
K11 Third issue DAF car transporter Yellow cab, yellow car transporter, rear loader, orange red lower chassis of car trailer, decals, metal plastic hubs, super king wheels, silver grill, headlights, white interior, 15/-d, Made 1971	£7	£5	£3	£2
3a Lemon cab, orange, yellow trailer, DAF decals, Made 1972	£7	£5	£3	£2
3b Mustard cab, mustard top part of car transporter, orange lower part of transporter, red grill, off white interior, DAF decals; Made 1973	£6	£4	£2	£1
3c Rich lemon cab, car transporter top, bright orange lower part of car transporter, cab grill, silver hubs, black super king tyres, black, silver hydraulics, decals, Made 1974/75	£5	£4	£2	£1
K11 Fourth issue AA pickup truck Golden yellow body, chassis, 6 super king wheels, silver headlights, grill, black bumpers, two white tow bars, red tow hooks, red flash on roof of cab, off white interior, orange tinted windows, AA decals, black motor decal on hood, door sides, red plastic tow hook, black support for two bars, hooks, 127 mm, 35-1; Made 1976				
K12 First issue Heavy breakdown wreck truck Light green body, green tinted windows, silver grill, silver metal hubs, black plastic tyres, knobby treads, yellow crane, black plastic cable, metal hook, BP shield decals, Matchbox Service Station decals on white background on sides of truck, decals on front of cab, 121 mm, 5/6d, 62-1, Made 1963	£10	£7	£5	£3
1a Dark green body, red plastic hubs, silver hook, cable on yellow crane, no background to decals, one shield BP decal in white square, 6/5d, Made 1966/67/68	£8	£6	£4	£2

	Mint boxed	Mint unbxd	Good cond	Fair cond
Flaw: Partition on top of cab for BP decal missing completely and part of roof at one side almost disappears into cab body	£50	£25		
K12 Second issue Scammell Crane truck Golden yellow body, clear windows in cab, green tinted windows in crane cab, off white interior, metal headlights, bumper, half base, off white plastic cable, grey metal hook, red hydraulic sleeves, decal, red plastic hubs, black plastic tyres, 9/6d, 153 mm, 63-1; Made 1970	£10	£7	£5	£3
2a White, orange body, orange chassis, crane, silver hook, green tinted windows on crane, amber tinted windows on cab, decals, silver hubs, headlights, bumpers, silver hook, Made 1971	£9	£6	£4	£2
2b Deep orange body, Made 1974	£9	£6	£4	£2
K12 Third issue Hercules mobile crane Yellow body, crane, red tinted windows on cab of lorry, black plastic supports, grey plastic tyres, metal hubs, black hydraulic stays, black plastic rod inside crane body, red hook; silver grill, headlights, red flashes on roof, decals, Made 1975	£1			
3a Orange tinted windows on cab, Made 1976				
K13 First issue Readymix concrete truck Orange, green tinted windows, silver grill, headlights, bumper, black base and rear pan; decals on doors, metal hubs, black plastic tyres, 114 mm, 58-1, 0 scale, 5/11d, Made 1963	£10	£7	£5	£3
1a Dark orange body, dark orange hubs, Made 1967	£10	£7	£5	£3
Flaw: Part of barrel missing at rear, no paint on remainder, bad factory fault	£50	£25		
K13 Second issue Building transporter Metallic green gold body, chassis, clear windows, metal grill, half base, metal gas tanks, tool box, black hubs, super king wheels, off white interior, red plastic supports on bed holding yellow building sections, 130 mm, 66-1, Made 1971	£3	£2	£1	50p
2a Light metallic gold, red tool box, petrol tank, Made 1973/74	£3	£2	£1	50p

88 LESNEY — KINGSIZE SPECIALS

	Mint boxed	Mint unbxd	Good cond	Fair cond
K13 Third issue DAF building transporter Light metallic green body, red tool box, petrol tank, red grill, Made 1975	£3	£2	£1	50p
K13 Fourth issue Aircraft transporter Red cab, chassis, off white interior, black supports holding white aircraft model, decals on sides, wings; black decal 101 on side of plane, white X4 decal on roof of cab on black square, silver hubs; black super king wheels, 203 mm, 38-1, Made 1976				
K14 First issue Jumbo crane Yellow greenish tinted body, green tinted windows, moveable crane, black hydraulic sleeve, black hook, metal hubs, black plastic wheels, no decal on first model, 6/-d; 125 mm, 58-1, O scale, Made 1963	£25	£15	£10	£5
1a Yellow; green tinted windows, moveable crane, black hydraulic sleeves, off white plastic cable, metal hook, decal on front bumper 127 EXH, red, white stripes, decal at rear of crane, red plastic hubs, black plastic tyres, Made 1964	£10	£7	£5	£3
Flaw: Very short stem on crane, only a ball instead of plastic hook	£25	£10		
1b Yellow crane, red box, decals	£7	£5	£3	£2
1c Yellow crane, red box, orange, white sticker decals	£7	£5	£3	£2
K14 Second issue Freight truck Metallic blue cab, chassis, sliding doors, red metallic box section, white roof, box base, clear windows, metal headlights, bumper, half base, blue axle cover, sticker decals on box sides, super king wheels, 14/-d, 140 mm, 64-1, Made 1971	£10	£7	£5	£3
K14 Third issue Freight truck Dark blue body, very inferior to previous models in all respects, 139 mm, Made 1976				
K15 First issue Merryweather fire engine Red, silver grill, headlights, bumper, clear windows, 2 gold bells, control box, grey metal ladder, decals, 143 mm, 96-1, TT scale, Made 1964	£25	£15	£10	£5
1a Plastic ladder, No 15 decals on door, shield, decals, 7/6d, Made 1965	£9	£7	£5	£3

	Mint boxed	Mint unbxd	Good cond	Fair cond
K15 Second issue The Londoner double decker bus Red, opening doors, super king wheels, decals, 121 mm, 92-1, Made 1974/75/76				
K16 First issue Dodge tipper tractor with twin tippers Metallic green, yellow tippers, green tow hook, black metal tow bar, green tinted windows, silver headlights, bumpers, exhaust, horns, dodge trucks in red letters on white background decals, 302 mm, 64-1, 14/11d; Made 1966, orange plastic hubs, 26 wheels	£25	£15	£10	£7
1a Red hubs, black plastic tyres, missing decals, Made 1967	£50	£25		
1b Red hubs, black plastic tyres, decals, deep green body, Made 1968	£10	£9	£7	£5
1c Red hubs, black plastic tyres, decals, Made 1969	£9	£8	£6	£4
1d Light yellow, blue tippers; super king wheels, Made 1971	£8	£7	£5	£3
K16 Second issue Petrol tanker Metallic red cab, silver grey metallic chassis, white tanker top, metallic red tanker chassis, decals, blue licence plate at rear, blue tinted windows, dark blue grill, dark blue ladders, red plastic pipe holders, 229 mm, 62-1, Made 1974/75/76				
K17 First issue Ford tractor & Dyson low loader & case bulldozer Green, no dome light on cab; metal grill, base, headlights, spare tyre mounted on top of loader front section, red bulldozer, yellow blade, canopy, green tracks, red hubs, black tyres, 242 mm, 54-1, Made 1966, Exhibition only, very rare	£250	£100		
K17 Second issue Ford loader with bulldozer Green, yellow gas tanks, connector bar, green tinted windows, red dome light, metal grill, base, decals on doors, green loader bed, spare tyre mounted on top section, red bulldozer, yellow motor blade, canopy, red plastic hubs; black plastic tyres, 15/6d, Made 1967	£10	£9	£7	£5
2a decals rear of cab, rear of tractor, decals on doors;	£7	£6	£3	£2
2b Decals, Please note connector base comes in a variety of colours, green or clear plastic, base of cabs comes with and without the number K17 with patent applied or patent number	£7	£6	£3	£2

LESNEY — KINGSIZE SPECIALS

	Mint boxed	Mint unbxd	Good cond	Fair cond
K17 Third issue Low loader with bulldozer Fluorescent green cab, chassis, loader, tractor, yellow blade, hood, decals on doors, dark green tracks, super king wheels, Made 1971	£6	£5	£3	£2
3a Fluorescent red cab, chassis, lime green loader, bed riveted, green tinted windows, red dome light, metal grill, base; tow slot, orange tractor, light yellow motor, blade, canopy, black rollers, decals, 242 mm, 54-1, Made 1972	£6	£5	£3	£2
K17 Fourth issue Articulated container truck Red metallic cab, silver grey chassis, white petrol tank, silver bumper, yellow grill, dark silver grey chassis transporter, red metallic mudguards, light blue rollers, 2 white box containers with blue roofs, decals, white interior of cab; 250 mm; 62-1, Made 1974	£7	£5	£4	£3
4a Deep orange box containers, decals	£5	£4	£3	£2
K17 Fifth issue Articulated container truck Red metallic cab, silver grey chassis, metallic red chassis on container, 2 white box containers with blue roofs, opening blue doors at rear, decals, off white cab interior; yellow grill, 251 mm; 62-1 Made 1976				
K18 First issue Articulated horsebox Green cab; metal grill, bumper, base; air horns on roof; green tinted windows, dark tan trailer, clear windows, opening doors, 4 plastic horses, 2 brown, 2 grey, no decals, red plastic hubs, black plastic tyres, 166 mm, 63-1; 10/6d, Made 1966	£50	£25	£15	£10
K18 Second issue Articulated horsebox Red cab, metal grill, bumper, base, air horns on roof, green tinted windows, light tan trailer, clear windows, 4 white plastic horses, decals, red hubs, opening doors, 167 mm, Made 1967	£7	£5	£4	£2
2a Red, tan, red hubs, black plastic tyres, rough base plate on cab, grey plastic interior, Made 1968	£7	£5	£4	£2
2b Red, tan hubs, rough base plate, green plastic interior, decals, doors have square holes at top, much heavier blocks, Made 1969	£7	£5	£4	£2
2c Red, dark tan horsebox, greenish plastic interior, dark grey horses; Made 1970	£7	£5	£4	£2
2d Red cab, bright orange trailer, white horses; off white interior, super king wheels; 14/-d; Made 1971	£5	£4	£2	£1
K18 Third issue Articulated tipper truck Metallic red cab, silver grey chassis, dark yellow body tipper wagon, dark blue tipper chassis, silver bumper, petrol tank, super king wheels, silver hubs; decals, 204 mm, 62-1, Made 1974	£5	£4	£3	£2
K18 Fourth issue Articulated tipper truck Silver grey cab, silver grey tipper wagon body, dark red chassis, super king wheels, decals, 203 mm, 62-1, Made 1976				
K19 First issue Scammell tipper truck Red cab, chassis, orange tipper; blue tinted windows, metal bumper, base, black grill, metal horns on roof of cab, decals, black plastic tyres, deep red plastic hubs, 6/11d, 121 mm, 69-1, Made 1968	£9	£7	£5	£3
1a Yellow tipper body	£5	£4	£2	£1
1b Green tinted windows, silver horns, silver trim, Made 1969/70	£5	£4	£2	£1
1c Super king wheels, Made 1971	£5	£4	£2	£1
1d Light yellow tipper body; 121 mm, Made 1975/76				
K20 First issue Ford transporter and tractors Red cab, trailer, green tinted windows, red dome light, metal grill, base, spare tyre on left side under trailer, yellow plastic tractor holders, 3 blue tractors, black steering wheels, yellow hubs, red plastic hubs on cab body, trailer, 15/6d, 228 mm; 62-1, Made 1968	£15	£10	£7	£5
1a Deep red body, trailer, metal side tanks, Made 1969	£25	£15	£10	£5
1b Yellow side tanks, Made 1969	£9	£7	£5	£3
1c Yellow trailer, red side tanks, super king wheels, Made 1971/72	£9	£7	£5	£3
K20 Second issue Cargo hauler & pallet loader Green metallic body, yellow petrol tank, tool box, super king wheels, metal hubs, 4 yellow plastic barrels on red plastic stand, 2 red plastic ramps at rear, off white cab interior, black metal grill, 191 mm, 66-1, Made 1973	£4	£3	£2	£1

LESNEY – KINGSIZE SPECIALS

	Mint boxed	Mint unbxd	Good cond	Fair cond
K20 Third issue Cargo hauler & pallet loader Light green body, chassis, yellow plastic tool box, petrol tank, yellow plastic grill, 4 yellow plastic barrels on red stand, 2 large tyres on red plastic holder, light pink plastic loading ramp, 190 mm, Made 1975/76				
K21 First issue Mercury Cougar Metallic blue, golden interior, silver bumper, grill, headlights, silver hubs, black plastic wheels, true guide steering, suspension, dark blue steering wheel, opening doors, 105 mm, 40-1, Made 1968	£25	£15	£10	£5
1a Metallic gold body, silver trim, off white interior, opening doors, 6/2d; Made 1969	£7	£5	£3	£2
1b Dark gold metallic body	£7	£5	£3	£2
K21 Second issue Cougar dragster Pink body, white interior, tow hook, silver motor, metal grill, headlights, bumpers, base, decals, small front wheels, large rear wheels, fast speed king type, 8/-d, 105 mm; 40-1, Made 1971	£5	£4	£2	£1
Flaw: **2a** Pink model, no decals	£15	£10		
2b Dark pink metallic	£4	£3	£2	£1
K21 Third issue Tractor transporter Blue metallic body, yellow grill, yellow tractor support, yellow ramp, 2 plastic purple tractors, silver engine, side flashes, super king wheels on all vehicles, 162 mm, 55-1, Made 1975/76				
K22 First issue Dodge charger Metallic dark blue cab, cream interior, tow hook, clear windows, metal grill, headlights, bumpers, base; blue tinted hubs, black plastic tyres, 6/2d, 115 mm, 48-1, Made 1969	£5	£4	£2	£1
1a Light blue interior; silver plastic hubs, Made 1970	£5	£4	£2	£1
K22 Second issue Dodge dragster Orange body; white interior, tow hook, silver mounted motor, silver headlights, bumpers, grill, super king wheels, decals, small front wheels, large rear wheels, 8/-d, 115 mm; 48-1, Made 1971	£4	£3	£2	£1
2a Bright metallic orange body, large decals, opening doors, white interior, Made 1972/73	£4	£3	£2	£1

	Mint boxed	Mint unbxd	Good cond	Fair cond
K22 Third issue Hovercraft Metallic blue, off white, 8 side windows, silver radar scanner, plastic propeller, decals, 127 mm, 116-1, Made 1975/76				
K23 First issue Mercury police car Silver grey; white interior, blue warning lights, silver horns, silver headlights, bumpers, silver hubs, decals, opening doors; 117 mm, 50-1, Made 1969	£5	£4	£2	£1
1a White, red interior, tow hook, Made 1970	£5	£4	£2	£1
1b White, super king wheels; Made 1971	£5	£4	£2	£1
K23 Second issue Articulated low loader with bulldozer Metallic blue cab, silver blue chassis, 4 super king wheels, yellow interior, yellow radiator, metallic gold ramp for joining trailer which also is metallic gold, red tractor, black tracks, yellow blade, silver trim, silver motor 241 mm, 62-1, Made 1976				
K24 First issue Lamborghini Miura Golden yellow; red interior, red driving wheel, silver headlights, bumper, grill, hubs, 8/-d, 102 mm, 43-1, Made 1969	£25	£15	£10	£5
1a Metallic red, black rear louvre on rear, lifts up, off white interior, metal grill, headlights, base, super king wheels, Made 1971	£5	£4	£3	£2
1b Metallic light red orange, speed king wheels, Made 1972/73	£5	£4	£3	£2
K24 Second issue Container truck Metallic red cab, chassis, white box container, red plastic roof, red plastic doors, off white interior, decals, 133 mm, 64-1, Made 1976				
K25 First issue Power boat and trailer Orange deck, black seats, white dash, clear windshield, white hull, motor, decals, brown plastic crib on yellow metal trailer, black axle covers, 7/-d, 153 mm, 43-1, Made 1971	£2	£1		
1a White plastic crib, dark orange body cover on boat, yellow trailer, Made 1972	£2	£1		
K25 Second issue Seaburst power boat and trailer Orange, white plastic boat, yellow plastic trailer, Made 1976				

LESNEY — KINGSIZE SPECIALS

	Mint boxed	Mint unbxd	Good cond	Fair cond
K26 First issue Mercedes ambulance White, white interior, blue dome light, red cross sticker decals; metal grill, base, speed king wheels, opening doors, stretcher case, silver headlights, bumpers, radiator, 9/-d, 105 mm, 45-1, Made 1971	£5	£4	£2	£1
1a Decals, Made 1974/75/76	£5	£4	£2	£1
1b Lighter plastic body, no trim. Made 1976				
K27 First issue Camping cruiser Bright yellow body, orange lift up, silver plastic headlights, bumpers, grill, white interior, silver stove, sink, red rear section, no decals, 117 mm, 48-1, Made 1971	£10	£7	£5	£3
1a decals with red dots, 10/-d, Made 1971	£5	£4	£2	£1
1b Pale yellow, super wheels, orange lift, Made 1975/76				
K28 First issue Juego drag sport Containing K21 Cougar mounted on yellow trailer, pulled by green Mercury station wagon, 19/-d, 250 mm, 50-1, Made 1971	£4	£3	£2	£1
K28 Second issue Drag pack K21 Cougar mounted on deep yellow golden trailer, pulled by dark green Mercury station wagon, Made 1975/76				
K29 First issue Miura seaburst set Contains K25 boat, trailer, pulled by K24 Lamborghini Miura, 250 mm, 43-1, same model released in 1976, Made 1971/72				
K30 First issue Mercedes C111 Metallic gold body, off white interior, silver headlights, black decals, opening hood revealing silver motor, 45p, 102 mm, 44-1, Made 1972	£2	£1		
1a Dark metallic blue, white flashes, opening bonnet, white interior, white plastic base, super king wheels, Made 1976				
K31 First issue Bertone Runabout Flame orange; opening engine compartment, decals, 102 mm; 44-1; Made 1972; still current 1976				
K32 First issue Shovel nose Yellow, No 2 decal on opening hood, super king wheels, 102 mm, 44-1, Made 1972	£1	50p		
1a Deep yellow, No 4 decal in orange shield, Made 1976				

	Mint boxed	Mint unbxd	Good cond	Fair cond
K33 First issue Citroen SM Red, rich ruby, yellow interior, green tinted windows, silver headlights, bumpers, opening doors, 115 mm, 43-1, Made 1972	£2	£1		
1a Dark brown body, golden yellow interior, white plastic base, super king wheels, 114 mm, Made 1976				
K34 First issue Thunderclap Yellow racing body, wide racing wheels, silver motor, exhaust, hubs, white driver, red helmet, No 34 decal, 108 mm, 40-1, Made 1972	£2	£1		
1a Black body, No 1 decal, white driver, white helmet, wide racing wheels, 108 mm, 40-1, Made 1976				
K35 First issue Lightning Red, no 35 decal, decals, silver plastic motor, base, wide racing wheels, 108 mm, Made 1972, 40-1	£2	£1		
1a White body, 35 decal, silver motor, white driver, wide racing wheels, 108 mm, Made 1976				
K36 First issue Bandolero Light blue, silver motor, base, white interior, orange tinted window, opening hood, wide racing wheels, red base, 108 mm, 41-1, Made 1973	75p	50p		
1a Dark metallic blue body, silver plastic base, 114 mm, Made 1976				
K37 First issue Sandcat Pink, green dappled, stippled effect on body, black hood, silver motor, silver trim, wide racing wheels, 91 mm, 38-1, Made 1973	£1	50p		
1a Bright orange; yellow dapple, silver motor, exhausts, black hood, black interior, wide racing wheels, 86 mm, Made 1976				
K38 First issue Gus's Gulper Pink, lift off body, mustard yellow interior, seats, silver motor, decals, super king wheels, still current model in 1976, 121 mm, 45-1, Made 1973				
K39 First issue Milligans Mill Green, lift off body, red seats, interior, red orange decals, silver motor, super king wheels, current issue darker green 1976, 108 mm, 45-1, Made 1973				

LESNEY — KINGSIZE SPECIALS

	Mint boxed	Mint unbxd	Good cond	Fair cond

K40 First issue Blaze trailer Deep pink, lift off hood, deep orange interior, seats, blue roof light, decals, lighter pink body in current model made 1976, 102 mm, 44-1, Made 1973

K41 First issue Fuzz buggy White, lift off roof, door, clear plastic window, red border on window, silver, black motor, super king wheels, still current 1976, 115 mm, 41-1, Made 1973

K42 First issue Nissan 270X Orange, clear plastic lift off hood, lemon interior, silver motor, No 8 decals, still current 1976; 102 mm; 44-1, Made 1973

K43 First issue Cambuster Yellow, silver motor, No 8 decal, silver motor, exhausts, hubs, super king wheels, no decals on sides, 105 mm, 45-1, Made 1973   £1

1a Large cambuster decal on doors, green, red, white, blue starred background, Made 1976

K44 First issue Bazooka Red, silver motor, exhausts, hubs, super king wheels at rear, small super wheels, black tyres at front, No 3 decal, current model 1976 with lighter body colour, 105 mm, 45-1, Made 1973

K45 First issue Marauder Pink, silver motor, No 7 decals, white driver, red helmet, wide racing king size super wheels, current model 1976, lighter pink body, white driver, yellow coated body, silver trim, 105 mm, 40-1

K46 First issue racing car pack Yellow estate car, 2 spare wheels on roof, red pack, red hubs, decals, silver plastic radiator, headlights, bumpers, orange interior; super king wheels, blue plastic trailer, Lightning racing car K34, yellow hook, tow bar on trailer, 261 mm, 50-1, Made 1973   £2   £1

1a Estate car, spare wheels, accs. on roof; deep orange body, No 34 decals, Made 1976

K47 First issue Easy rider Silver forks, blue plastic seat; backrest, yellow driver, red motor, silver headlight, exhausts, spoked silver plastic wheels, still current 1976; 121 mm, 20-1, Made 1973

K48 First issue Mercedes 350 SLC Red, yellow interior, silver headlights, bumpers, grill, super king wheels, opening bonnet, 105 mm, 45-1, Made 1973   £2   £1

1a Bright red body, orange interior, silver motor, Made 1976

K49 First issue Ambulance White, red flashes on doors, shield decals, yellow plastic roof rack, red flash light, silver headlights, bumpers, grill, horns, super king wheels, Made 1973   £2   £1

1a Black lettered decal AMBULANCE on doors, red cross decals on hood, side rear windows, opening back door, two plastic stretcher bearers, red stretcher, orange flashes, blue flasher, silver plastic headlights, bumpers, grill, bright red roof rack, super king wheels, Made 1976

K50 First issue Street rod Bright yellow, no decals, silver motor, super king wheels, black seats, silver headlights, grill, bumpers, silver exhausts, black plastic hood, 79 mm, 45-1, Made 1973   £5   £4   £3   £2

1a Orange body, green chassis, silver headlights, grill, bumpers, tool box cover at rear, black decal '8' on sides, silver motor, exhausts, Made 1973/74   £5   £4   £3   £2

K50 Second issue Street rod Yellow orange body, bright orange dot decals on doors, silver plastic rear box, green chassis, silver headlights, bumpers, grill, motor, exhausts, Made 1976

K51 First issue Barracuda Deep metallic blue, white driver, red helmet, silver motor, No 5 decal, super king wheels, silver grid above driver's head, 108 mm, 40-1, Made 1973   £2   £1

1a Pale blue body, coloured decals, yellow seat, white driver, no helmet, silver motor, cover above driver, super king wheels, 108 mm, Made 1976

K52 First issue Datsun rally car Golden yellow, 3 silver headlights, bumper bar, radiator frame, rear bumper bar, hub caps, wide super king wheels, No 52 decal, 105 mm. 40-1. Made 1974/75   £2   £1   50p

1a Bright orange body, decals lighter blue, doors, opening bonnet, silver motor, headlights, bumpers, grill; super king wheels, Made 1976

K53 First issue Hot fire engine Red metallic, 2 firemen, black with gold helmets, silver motor, headlights, radiator frame, exhausts, rear supports, yellow plastic ladder; decals; silver bell; super king wheels; 98 mm; 45-1; Made 1976

K54 First issue AMX javelin Metallic pink, orange interior; green tinted windows; silver headlights, bumpers; yellow base, plastic tow hook, No 7 decals, super king wheels, 108 mm, 45-1, Made 1976

K55 First issue Corvette caper cart Deep metallic blue; No 55 decals, wide super king wheels, yellow interior; 108 mm, 43-1, Made 1976

K56 First issue Maserati Bora Metallic silver grey, decals, yellow interior, black roof hood, silver headlights, grill, bumpers, silver motor, super king wheels, 101 mm, 43-1, Made 1976

K57 First issue Javeling drag racing set Consisting of K54 AMX Javelin, K39 Milligans mill racer, yellow plastic trailer, Please note colours may vary in cars that are in these sets, 247 mm, 45-1, Made 1976

K58 First issue Corvette power boat set Consisting of K55 Corvette caper cart, K25 Seaburst power boat, yellow plastic trailer, Please note colours may vary in all models in these sets, 257 mm, Made 1976

K59 First issue Ford Capri Mk II Metallic silver, silver headlights, bumpers, exhaust, yellow, black decals, super fast wheels, 105 mm, 41-1, Made 1976

K60 First issue Ford Mustang Metallic red, silver mount on bonnet, silver headlights, grill, red bumpers, silver exhausts, super king wheels, 107 mm, 45-1, Made 1976

K61 First issue Police car White, silver horns, blue flasher on roof, yellow flash lights at side of radiator, silver bumpers, radiator grill, hubs, police decals, speed king wheels, 107 mm, 45-1, Made 1976

Major packs

A complete collection of models that were made between 1960 and 1966

	Mint boxed	Mint unbxd	Good cond	Fair cond
M1 First issue Caterpillar earth remover Yellow, yellow driver, 4 black large tyres, 2 small black tyres, metal hubs; round metal axles, 116 mm, 125-1, Made 1960	£30	£25	£15	£10
1a Silver plastic hubs, tan body, tan driver, Made 1961	£15	£10	£7	£5
M1 Second issue BP Autotanker Lime green, cream, decals, 4 metal axles, 8 metal wheels, 102 mm, 89-1, HO scale, Made 1962	£15	£10	£5	£3
Flaw: Shield decals missing	£50	£25		
2a Black plastic wheels, green tinted windows, 102 mm, 89-1, Made 1963	£10	£7	£5	£3
2b Dark yellow tinted roof, light chassis, lower part of body green, Made 1964	£10	£7	£5	£3
M2 First issue Bedford articulated Walls ice cream lorry White, decals, royal blue cab, silver headlights, bumper, grill, grey metal wheels, round axles, 101 mm, 95-1, TT scale, Made 1960	£30	£25	£10	£7
1a Dark maroon flash decals, dark blue cab, grey metal wheels, Made 1961	£30	£25	£10	£7
M2 Second issue Davies tyre truck Red, grey bumper, grill, red chassis, red rear opening doors, 6 black plastic wheels, decals, 117 mm, 77-1, 00 scale, Made 1962	£15	£10	£7	£5
2a Darker red cab, chassis, darker red doors, dark decals, 117 mm, 00 scale, Made 1963	£15	£10	£7	£5

LESNEY — MAJOR PACKS

	Mint boxed	Mint unbxd	Good cond	Fair cond
2b Dark green box container, Made 1964	£12	£9	£4	£2
M2 Third issue Articulated box truck Silver grey, brownish red container truck body, dark brown doors at rear, decals, 117 mm, 77-1, 00 scale, Made 1965	£15	£10	£7	£5
3a Silver grey, blue tinted windows, maroon body, black rear doors, black plastic wheels, decals, Made 1966	£20	£15	£10	£5
M3 First issue Thorneycroft antar with Sankey 50 ton tank transporter and centurion Mk III tank Olive green; 10 metal wheels, grey, grey tracks on tank, 10 wheel transporter, 156 mm, 125-1, Made 1960	£30	£25	£10	£7
1a Black plastic wheels, 155 mm, Made 1964	£25	£15	£7	£5
M4 First issue Ruston Bucyrus model 22 RB excavator Maroon, yellow shovel, arm, black chassis, green tracks, decals, metal rollers, very rare first model, 921 mm, HO-TT scale, Made 1960	£30	£25	£15	£10
M4 Second issue Dragline excavator Dark maroon cab, black chassis, green tracks, dark yellow shovel, sleeve, base, decals; 82 mm, 92-1, TT scale, Made 1961	£10	£7	£5	£3
2a Light yellow shovel, sleeve, base, Made 1962	£10	£7	£5	£3
2b Deep chocolate cab, dark tan sleeve, digger, base, white plastic rollers, green tracks, 98 mm, Made 1963	£9	£6	£4	£2
M4 Third issue GMC Tractor with hopper train Red, tow bars, silver grey grill, headlights, bumpers, silver grey body on hopper wagons, red plastic hubs, grey plastic tyres, 2 spare wheels, 286 mm, 67-1, Made 1965	£25	£15	£10	£5
3a Dark grey hopper wagons, dark maroon cab, decals on hopper wagons, Made 1966	£20	£10	£7	£5
M5 First issue Massey Ferguson combine harvester Red, tan driver, dark grey metal wheels, large black plastic tyres front, small plastic tyres at rear, orange blades, decals, 58:1, 0 scale, Made 1960	£25	£15	£10	£5
1a Grey plastic hubs, Made 1961	£15	£9	£7	£5
Flaw: Missing driver, missing steering wheel, short steering column, short drivers seat position, missing stack	£50	£25		
M5 Second issue Massey Ferguson combine harvester Deep maroon body, dark brown rotating blades on shove, green hubs, 117 mm, 58-1, Made 1964	£12	£10	£6	£4
2a Yellow driver, orange hubs, Made 1965	£10	£7	£5	£3
2b Red hubs, large black plastic wheels, small front wheels, Made 1966	£10	£7	£5	£3
M6 First issue Scammell tractor and transporter Deep blue, dark red, body, chassis, decals, 18 grey metal wheels, 9 round axles, 299 mm, TT scale, 116-1, Made 1960	£30	£25	£15	£10
1a 18 plastic wheels, grey, Made 1962	£15	£10	£7	£5
M6 Second issue Racing car transporter Green, off white interior, green tinted windows, decals, orange hubs, black plastic wheels, 127 mm, 54-1, Made 1965/66	£15	£10	£7	£5
M7 First issue Ford Thames cattle truck Red; long tan cattle box body, silver grill, bumpers, headlights; grey metal wheels, 81:1, 00 scale, Made 1960	£25	£15	£9	£7
1a Dark red cab; dark tan truck, silver grey wheels, Made 1965	£15	£10	£7	£5
M8 First issue Mobilgas petrol tanker Red, silver headlights, grill, bumper bar; 6 grey metal wheels, 3 round axles, decals, 92:1, HO TT scale, Made 1960	£15	£10	£7	£5
1a Dark maroon body, silver grey plastic wheels, Made 1963	£15	£10	£7	£5
M8 Second issue Car transporter Light blue cab, light blue ramp, orange transporter, decals, silver headlights, bumpers, blue grill, black hubs, black tyres, 209 mm, 73-1, 00 scale, Made 1964	£15	£9	£7	£5
2a Green cab, deep orange transporter, decals, grey plastic tyres, orange hubs, Made 1966	£10	£7	£5	£3
M9 First issue Interstate double freighter Dark purple cab, trailer tow bars, opening doors at rear, silver horns, grid, headlights, bumper bars; off white interior, decals, black plastic wheels, 34 in all, 289 mm, 100-1, TT scale, Made 1962	£25	£15	£10	£5

1a Light blue cab, silver freight cars, blue, grey plastic wheels, darkish orange background decals, golden

	Mint boxed	Mint unbxd	Good cond	Fair cond
horns on cab roof, 288 mm, Made 1963	£25	£15	£10	£5
Flaw The arrows pointing towards the rear	£100	£50		
1b Light blue cab, green silver tinted freight cars, Made 1964/65	£15	£10	£7	£5
1c Dark royal blue cab, golden horns, silver blue freight line cars, 34 black plastic wheels, decals, Made 1966	£25	£15	£10	£5
M10 First issue Whitlock dinkum dumper Dark tan body, white plastic hubs, silver grill, exhaust, headlights, licence plate, black tyres, 108 mm, 70-1, 00 scale, Made 1962	£15	£10	£7	£5
1a Light body, grey tyres, 108 mm, Made 1963	£10	£7	£5	£3
1b Light yellow body, red hubs, Made 1966	£9	£7	£5	£3

In 1964 an addition to the major pack issue was advertised. This was Fruehauf Hopper Train, but although a few prototype models were made the actual model never reached the shops. A late change was made and another G.M.C. Tractor with Hopper Train was put out in the place of the original.

M4 Fruehauf Hopper train Green cab, chassis, silver grill, headlights, bumper bar, clear windows, dark steering wheel, decals, metal hubs on cab, hopper wagons, 18 plastic black tyres, knobby treads, two rust coloured hopper wagons made 1965 but never released, very rare model, perhaps the rarest in the history of Lesney era, 279 mm, 67-1,

£5,000 £1,000

Gift sets

In 1961 Lesney introduced some beautifully designed boxes and started a gift set series. They are now very rare.

G1 First issue Commercial vehicles Contents Nos. 5, 10,20,37,47,46,51, 69, Made 1961	£45	£40	£20	£10
G1 Second issue Commercial vehicles Contents Nos. 5,10,12,13,14,21,46, 74, Made 1962	£50	£45	£25	£15
G1 Third issue Motorway set Contents Nos. 6,10,13, 33,34,38,48,55,71, Roadway Road one, BP road signs, Made 1964	£40	£35	£15	£10
G1 Fourth issue Service station set Contents Nos, MG1 13,31,64, Accessory pack No A1, Made 1966	£25	£20	£15	£10
G1 Fifth issue Service station set Made 1968	£10	£7	£5	£3
G1 Sixth issue Service station set 28/6d, Made 1970	£10	£7	£5	£3
G1 Seventh issue Service station set Super fast wheels, £1.80, Made 1972	£9	£7	£5	£3
G1 Eighth issue Twin launcher set Made 1976				
G2 First issue Car transporter & load Contents Nos. 22,25,33,39,57,75, Accessory pack A2; Made 1961	£30	£25	£10	£7
G2 Second issue Car transporter set Contents Nos. 25,30,31,39,48,65, A2, Made 1962	£30	£25	£10	£7
G2 Third issue Transporter set Contents Nos. 28,32, 44,53, M8, Made 1964	£30	£25	£10	£7
G2 Fourth S issue Car transporter set Nos. 22,28,36, 75, M8, Made 1966	£30	£25	£10	£7
G2 Fifth issue Transporter set Five vehicles, 21/-d, Made 1968	£30	£25	£10	£5

LESNEY — GIFT SETS

	Mint boxed	Mint unbxd	Good cond	Fair cond
G2 Sixth issue Transporter set 22/6d, Made 1969	£30	£25	£10	£7
G2 Seventh issue Transport set 28/6d, Made 1970	£10	£9	£7	£5
G2 Eighth issue Transporter set 33/-d, Made 1971	£15	£10	£7	£5
G2 Ninth issue Transporter set £1.75, Made 1972	£12	£10	£7	£5
G2 Tenth issue Big mover set Made 1973	£10	£7	£5	£3
G2 Eleventh issue Big mover set Made 1974	£10	£7	£5	£3
G2 Twelfth issue Big mover set Made 1976				
G3 First issue Constructional plant Contents Nos. 2,6,15,16,18,24,28, M1, Made 1961	£40	£35	£15	£10
G3 Second issue Farm & agricultural set Contents King size nos K3, K11, M5, M7, Made 1963	£25	£15	£7	£5
G3 Third issue Vacation set Contents Nos 12,23,27, 42,45,48,56,68, 17/11d, Made 1966	£25	£15	£7	£5
G3 Fourth issue Vacation set Contents Nos 12,23,27, 42,45,46,68, sports boat with trailer, 17/11d, Made 1967	£25	£15	£7	£5
G3 Fifth issue Farm set Contents 8 vehicles, 19/6d, Made 1968	£15	£10	£7	£5
G3 Sixth issue Farm set 21/-d, Made 1969	£15	£10	£7	£5
G3 Seventh issue Superfast racing special 19/11d, Made 1970	£10	£7	£5	£3
G3 Eighth issue Superior superfast racing set 25/-d, 35 car transfers, Made 1971	£10	£7	£5	£3
G3 Ninth issue Wild ones set full colour attractive range, Made 1973	£5	£4	£3	£2
G3 Tenth issue The wilder ones set Made 1976				
G4 First issue Agricultural implements & farm vehicles Contents Nos. 12,23,31,35,50,72, M7, Made 1961	£40	£20	£10	£7
G4 Second issue Grand Prix race track set Contents Nos. 13,14,19,41,47,73,52,32, M4, Roadway R4, Made 1963/64	£30	£15	£7	£5
G4 Third issue Race track set Contents Nos. 13,19 green,19 orange,29,41 white,41 yellow,52 blue, 52 red,54, M6, Roadway R4, 30/-d, Made 1966	£30	£15	£7	£5
G4 Fourth issue Race 'n' rally set 24/-d, Made 1968/69	£25	£10	£7	£5
G4 Fifth issue Truck set 24/-d, Made 1970	£15	£9	£6	£4
G4 Sixth issue Truck set Superfast wheels, Made 1971	£15	£9	£6	£4
G4 Seventh issue Superfast champions set Made 1972	£10	£7	£5	£3
G4 Eighth issue Team matchbox set Made 1974	£10	£7	£5	£3
G4 Ninth issue Superfast champions set Made 1976				
G5 First issue Military vehicles set Contents Nos. 54, 62,63,64,67,68, M3, Made 1961	£50	£25	£15	£10
G5 Second issue Military vehicles set Contents Nos. 12,49,54,61,64,67, M3, Made 1964/65	£45	£20	£10	£7
G5 Third issue Fire station set Contents Nos. 29,54, 59, MF1, 21/-d, Made 1966/67	£25	£15	£9	£7
G5 Fourth issue Famous cars of yesteryear set 26/6d, Made 1968/69	£75	£60	£15	£10
G5 Fifth issue Famous cars of yesteryear set Made 1969	£75	£60	£15	£10
G5 Sixth issue Famous cars of yesteryear set £1.90, Made 1971/72	£75	£50	£15	£10
G5 Seventh issue Famous cars of yesteryear set Made 1973	£60	£50	£15	£10
G5 Eighth issue Famous cars of yesteryear set Made 1974/75/76				
G6 First issue Models of yesteryear Content Nos. Y1, Y2,Y5,Y10,Y13, Made 1961	£500	£300	£30	£25
G6 Second issue Models of yesteryear Content Nos. Y6, Y7, Y10, Y15, Y16, Made 1962	£250	£100	£30	£25
G6 Third issue Veteran cars set Contents yesteryear Nos. Y5, Y6, Y7, Y15, Y16, Made 1963	£350	£100	£30	£25
G6 Fourth issue Commercial truck set Contents Nos. 6,15,16,17,26,30,58,62, Made 1964	£40	£20	£10	£7
G6 Fifth issue Commercial truck set Containing Nos, 16, 17, 25, 26, 30, 69, 70, 71, 17/11d, Made 1966	£25	£15	£10	£7
G6 Sixth issue Truck set 19/6d, Made 1968	£20	£10	£7	£5
G6 Seventh issue Truck set 21/-d, Made 1969	£20	£10	£7	£5

LESNEY — GIFT SETS 97

	Mint boxed	Mint unbxd	Good cond	Fair cond
G6 Eighth issue Drag race set Super fast wheels, £1.25, Made 1972	£15	£9	£6	£4
G6 Ninth issue Drag race set Made 1973/74	£4	£3	£2	£1
G6 Tenth issue Drag race set Made 1976				
G7 First issue Models of yesteryear Content Nos Y3, Y8, Y9, Y12, Y14, Made 1961	£250	£100	£30	£25
G7 Second issue Models of yesteryear Content Nos Y3, Y4, Y11, Y12, Y13, Made 1962	£250	£100	£30	£25
G7 Third issue Models of yesteryear Content Nos Y3, Y4, Y11, Y12, Y13, Made 1963	£250	£100	£30	£25
G7 Fourth issue Veteran & Vintage set Content Nos Y2, Y5, Y10, Y15, Y16, Made 1964	£250	£100	£30	£25
G7 Fifth issue Models of yesteryear Content Nos Y1, Y3, Y11, Y14, 22/6d, Made 1966	£200	£100	£20	£15
G7 Sixth issue Ferryboat set £1.20, Made 1972	£6	£4	£3	£2
G7 Seventh issue Ferryboat set Made 1973	£5	£4	£3	£2
G7 Eighth issue Car Ferry set Made 1976				
G8 First issue Civil Engineering Construction set Kingsize Nos. K1, K2, K3, K5, K6, Made 1962	£40	£30	£15	£10
G8 Second issue Construction set King size Nos K1, K7, K10, K13, K14, Made 1964	£40	£30	£15	£10
G8 Third issue King Size set K1, K12, K11, K15, 26/11d, Made 1966	£30	£20	£12	£10
G8 Fourth issue Incomplete, no definite set for 1976				
G9 First issue Major Packs set Content Nos, M1, M2, M4, M6, Made 1962	£50	£25	£15	£10
G9 Second issue Service Station set Matchbox series Nos. MG1, 13, 33, 71, Accessory pack A1, Made 1964	£15	£9	£7	£5
G9 Third issue Commando Task Force set Made 1976				
G10 First issue Garage & Service station set Matchbox series MG1, 13, 25, 31, A1, Made 1962	£30	£15	£7	£5
G10 Second issue Garage gift set Made 1963	£25	£10	£7	£5
G10 Third issue Fire station set Matchbox series Nos. MF1, 14, 59, 2 models of No 9, Made 1964	£25	£15	£7	£5
G10 Fourth issue Thunder Jets set Made 1976				
G11 First issue Strike Force set 1976				
G12 First issue Rescue set Made 1976				
G13 First issue Construction Site set Made 1976				
G14 First issue Grand Prix set Made 1976				
G15 First issue Car Transporter set Made 1976				

Accessory packs

	Mint boxed	Mint unbxd	Good cond	Fair cond
AP1 First issue Esso Pumps & Forecourt sign Red pumps, decals, attendant in white, red face, legs, Large Esso sign, 00 scale, 75:1, Made 1960	£2	£1	50p	25p
AP1 Second issue Esso Pumps & Sign Red pumps, decals, man, cream body, face, red legs, 115 mm, 75-1, Made 1961	£3	£2	£1	50p
AP2 First issue Articulated 4 car Transporter truck Blue, no decals, black metal wheels, silver headlights, bumper bar, grill, 71-1, Made 1960	£50	£25	£15	£10
AP2 Second issue Articulated car Transporter truck Deep blue cab, light blue transporter, silver headlights, grill, bumper bar, dark grey metal wheels, decals, 71-1, Made 1961	£10	£7	£5	£3
AP2 Third issue Articulated Car Transporter Rich clear blue cab, body, silver headlights, bumper, grill, red decals on lower body, grey metal wheels, silver trim, 165 mm, 71-1, rare model, Made 1962	£25	£15	£7	£5
AP3 First issue Metal Lock Up Garage Orange brick side walls, red brick roof, green doors, 00 scale, 72-1, Made 1960	£2	£1	50p	25p
AP3 Second issue Metal Lock Up Garage Lemon brick side walls, dark maroon roof, olive lime green doors, 74 mm, 00 scale, Made 1961	£2	£1	50p	25p
AP3 Third issue Garage for Single Car Dark brown side walls, red roof, green doors, 76 mm, 72-1, 00 scale, Made 1962/63	£2	£1	50p	25p
AP4 First issue Road Signs Consisting of 8 road signs, made in 1960 ideal for railway and model town	£10	£4	£3	£2

LESNEY — GIFT SETS

	Mint boxed	Mint unbxd	Good cond	Fair cond
AP4 Second issue Road Signs Height 40 mm, 77-1, 00 scale, Made 1961	£7	£4	£2	£1
AP4 Third issue Road Signs Made 1962/63	£5	£3	£2	£1
AP5 First issue Double fronted store Pale blue door, grey main building, decal, window decorated, 73 mm, 00 scale, Made 1961	£10	£5	£3	£2
AP5 Second issue Double fronted store Darker colours, 72 mm, 72-1, 00 scale, Made 1962/63	£5	£3	£2	£1
AP6 No R1 First issue Main highway Complete, By-pass, link road, construction signs, 3 dimensional stand up model, Made 1960, This model comes in a packet instead of box, as do Nos 3,4,2 with R in front of series	£10	£3	£2	£1
AP6 No R1 Second issue Main highway Made 1961	£3	£2	£1	50p
AP6 Third issue Main highway No R1 Made 1963	£3	£2	£1	50p
AP7 No R2 First issue The Heart of London Piccadilly Circus, Eros, Regent Street, Haymarket, Piccadilly, famous streets, Made 1961	£10	£5	£2	£1
AP7 No R2 Second issue Heart of London Made 1963	£4	£3	£2	£1
AP8 No R3 First issue The Mall Consisting of Admiralty Arch, Buckingham Palace, St. James' Park, famous buildings, Made 1961	£10	£5	£3	£2
AP8 No R3 Second issue The Mall Made 1962/63	£5	£3	£2	£1
AP9 No R4 First issue Motor racing circuit 3 dimensonal cutouts of pits, grandstands, spectator figures, bridges, results boards, Made 1961	£10	£5	£2	£1
AP9 No R4 Second issue Motor racing circuit Made 1962/63	£5	£2	£1	50p
AP10 P1 First issue Painting Yesteryears Popular Lesney Matchbox models shown in action, wonderful true colours, including selection of models of Yester-year, Made 1961	£25	£5	£1	50p
AP11 P2 First issue Military & Motoring Matchbox series vehicles shown in a variety of situations, including good racing, military scenes; Made 1961	£25	£5	£1	50p
AP12 No P3 First issue The Farm scene Book set out with agricultural, farming models, full colour	£25	£5	£1	50p
AP13 No P4 First issue Commercial Greatness Fully Illustrated book with exciting scenes, commercial in the Matchbox range, Made 1961	£25	£5	£1	50p
AP14 PC-1 First issue Paint & Crayon set Complete with wide selection water paints, fine brush, colouring crayons, Made 1961	£25	£5	£1	50p
AP15 No MF-1 First issue Fire station Tough plastic, double doors, firemans pole, transfers, building grey, red, decals, 9/-d, Width 235 mm, height 80 mm, 75-1, 00 scale, Made 1962	£25	£5	£2	£1
AP16 No MG-1 First issue Service station Showrooms on 2 floors, forecourt, pumps, greasing ramp, decals, Garage green, and white, 9/-d, width 235 mm, height 112 mm, 75-1, 00 scale, Made 1963	£25	£5	£2	£1
AP17 No MG-1a First issue Service station Showrooms on 2 floors, forecourt, pumps, greasing ramp, decals, pump attendant, red, tan, white, clock, very rare, 9/-d, width 235 mm, height 112 mm, 75-1, 00 scale, Made 1962/63	£50	£10	£5	£3

Dinky Santa Claus 1960 series

Dinky and the magnificent sixty-seven

The firm of Dinky made their name famous with the craftsmanship that went into their models.

	Mint boxed	Mint unbxd	Good cond	Fair cond
No 100 Lady Penelope's FAB 1 Pink with lady herself and her driver.	£20	£15	£8	£4
No 101 Thunderbird 2 Green 143 mm	£15	£10	£3	£2
No 110 Aston Martin DB5 Red and black, opening bonnet 111 mm	£5	£4	£2	£1
No 113 MGB Sports Car White with driver, opening doors 86 mm	£5	£4	£2	£1
No 114 Triumph Spitfire Gold with lady driver, opening bonnet 89 mm	£5	£4	£2	£1
No 115 Plymouth Fury Sports Driver and passenger, pullout twin aerials 122mm	£5	£4	£3	£1
No 116 Volvo 1800 S Red with opening doors, opening boot and bonnet	£5	£4	£3	£2
No 127 Rolls Royce Silver Cloud Mark III Gold with opening doors and bonnet, approx 125 mm	£10	£7	£5	£2
No 128 Mercedes Benz 600 Red, opening bonnet and boot, all four doors open with chauffeur	£10	£8	£5	£3
No 129 Volkswagen 13 Sedan Blue, opening doors, boot and bonnet	£5	£4	£3	£2
No 130 Ford Consul Corsair Blue with opening bonnet	£5	£4	£3	£2
No 132 Ford 40 RV Silver grey, opening boot and bonnet	£6	£5	£3	£2
No 133 Ford Consul Cortina 102 mm	£5	£4	£3	£2

	Mint boxed	Mint unbxd	Good cond	Fair cond
No 135 Triumph 2000 Green body with white roof, open boot and bonnet, approx 120 mm	£5	£4	£3	£2
No 136 Vauxhall Viva Blue with opening bonnet and boot	£5	£4	£3	£2
No 138 Hillman Imp Red, opening bonnet and boot, 86 mm	£5	£4	£3	£2
No 140 Morris 1100 Saloon Dark blue or light blue, approx 102 mm	£4	£3	£2	£1
No 142 Jaguar Mark X Light blue, approx 127 mm	£5	£4	£3	£2
No 147 Cadillac 62 Green with red interior fins, approx 127 mm	£6	£5	£4	£2
No 151 Vauxhall Victor 101 Yellowy green, opening boot and bonnet	£5	£4	£3	£2
No 152 Rolls Royce Phantom V Lim Black with gold inside, opening boot and bonnet, approx 152 mm	£10	£8	£5	£4
No 153 Aston Martin DB 6 Blue opening boot and bonnet, opening doors	£6	£5	£4	£2
No 154 Ford Taunus Dark yellow with opening doors, boot and bonnet, approx 127 mm	£5	£4	£2	£1
No 156 Saab 96 Red, opening doors, 98 mm	£5	£4	£3	£1
No 158 Rolls Royce Silver Shadow Red, all doors open, opening boot and bonnet, approx 127 mm	£20	£15	£5	£2
No 161 Ford Mustang (Fastback 2+2) White, opening doors, boot and bonnet	£6	£5	£3	£1
No 162 Triumph 1300 Blue, opening boot and bonnet, approx 92 mm	£5	£4	£2	£1

DINKY – SELECTED SPECIALS

Triumph. Post Office van

	Mint boxed	Mint unbxd	Good cond	Fair cond
No 163 Volkswagen 1600 TL Fastback. Red, opening doors, boot and bonnet, 102 mm	£6	£5	£2	£1
No 164 Ford Zodiac. Silver grey, opening all doors, boot and bonnet, approx 114 mm	£10	£9	£5	£3
No 170 Lincoln Continental. Blue and white, opening bonnet and boot, approx 127 mm	£6	£5	£3	£2
No 171 Austin 1800. Blue with opening bonnet and boot, approx 89 mm	£5	£4	£2	£1
No 172 Fiat 2300 Station Wagon. White and dark blue roof, opening back and bonnet, approx 108 mm	£6	£5	£3	£2
No 181 Volkswagen. Blue, approx 76 mm	£5	£4	£3	£1
No 183 Morris Mini Minor (Automatic). Red, black roof and red body, opening doors and bonnet, approx 51 mm	£10	£7	£2	£1
No 196 Holden Special Sedan. Blue, white roof, opening boot and bonnet, approx 108 mm	£6	£5	£3	£2
No 197 Morris Mini Traveller. Green with brown trim, yellow interior, approx 73 mm	£6	£5	£3	£1
No 199 Austin 7 Countryman. Blue, brown trim, red interior, approx 73 mm	£6	£5	£3	£1
No 212 Ford Cortina rally car. Cream, light on roof, black bonnet, opening doors, decals 'Dunlop' on side, No 8 decal on door and bonnet, approx 102 mm	£6	£5	£4	£2
No 214 Hillman Imp rally car. Blue, white trim, red interior, opening bonnet and boot, '35' decal on doors and bonnet, 86 mm	£6	£5	£3	£2
No 215 Ford GT Racer. White, opening bonnet and back, black decal '7' on doors, 96 mm	£6	£5	£3	£2
No 216 Dino Ferrari. Red, opening doors and engine cover, 98 mm	£6	£5	£3	£2
No 524 Panhard 24 C. Slatey grey, red interior, approx 76 mm	£6	£5	£3	£2
No 796 Healey Sports boat on trailer. Green, off white, orange trailer, 108 mm	£3	£2	£1.50	£1
No 485 Santa special model T Ford. White and red, red Santa figure, tree and toy bag, Xmas decals on sides, 79 mm	£40	£30	£10	£5
No 475 Model T Ford. Blue, green driver with lady in white, gold trim with red seats, 79 mm	£50	£40	£10	£5
No 476 Bull Nosed Morris Oxford. Blue, yellow, dark blue, grey roof, driver, red and gold trim, 92 mm	£50	£40	£10	£5
No 125 Fun Ahoy Set. Ford Consul Corsair; light blue, driver Healey sports boat and trailer, cream and green, driver and passenger	£15	£10	£7	£5
No 117 4 Berth Caravan/Trailer. Yellow, transparent roof, opening door, tow hook, approx 127 mm	£6	£5	£3	£1
No 118 Tow Away Glider Set. Triumph 2000, cream and yellow, blue roof; glider, cream; trailer with Southdown Gliding Club decals on side, 289 mm	£15	£10	£5	£3
No 237 Mercedes Benz racing car. Silvery grey, driver, red wheels, red seats, red decal 30 on bonnet, 98 mm	£10	£7	£4	£2
No 240 Cooper racing car. Dark blue, white driver, orange hat, black and silver wheels, No 20 black decals on sides and bonnet, 83 mm	£5	£4	£2	£1
No 242 Ferrari racing car. Red, white driver, silver cap, black tyres, silver caps, No 36 on sides and bonnet, 89 mm	£5	£4	£2	£1
No 241 Lotus racing car. Green, white driver, silver cap, black tyres with silver hubs, No 24 on black decals on door and bonnet, 83 mm	£5	£4	£2	£1
No 243 BRM racing car. Green with gold cover, lift off white driver, silver cap, black tyres and silver hubs, No 7 decal on sides and bonnet, 83 mm	£5	£4	£2	£1

DINKY — SELECTED SPECIALS

	Mint boxed	Mint unbxd	Good cond	Fair cond
No 258 Cadillac USA Police Car. Black with white door, aerial, red roof light, number on boot, sides with police decal, 111 mm	£6	£5	£3	£2
No 264 Cadillac RCMP Patrol Car. Royal blue, white door, aerial, red roof light, crest of RCMP decal on door, 121 mm	£6	£5	£3	£2
No 263 Superior Ambulance Criterion. Red & cream (rare)	£25	£20	£8	£5
driver & mate, stretcher with patient, 127mm cream with red trim	£15	£10	£4	£2
No 278 Vauxhall Ambulance. Silver, driver, blue light on roof, red X on front door, ambulance decal on front roof, approx 83 mm	£10	£9	£5	£3
No 267 Super Superior Cadillac Ambulance. Cream, roof lights, wide red trims with words Ambulance on roof, sides and back, 152 mm	£15	£14	£8	£5
No 273 RAC Patrol Van. RAC colours, opening doors at rear, decals on sides and roof, 76 mm	£15	£10	£5	£3
No 274 AA Patrol Van. Yellow, opening rear doors, decals on side and front roof, all details perfect, 79 mm	£15	£10	£5	£3
No 275 Brinks Armoured Car. Grey and off white, dark blue bumper, opening rear doors with money bags and gold bars, gold trim sides, 'Brinks' decals on sides (security 1859), 121 mm	£15	£10	£5	£3
No 277 Superior Criterion Ambulance. Dark blue body, cream top, two drivers, white tyres, ambulance decals on sides of rear windows, flashing light, stretcher, patient, 127 mm	£15	£14	£10	£5
No 280 Midland Mobile Bank. Cream, blue, silver, gold trim, black tyres, silver hubs, decals on side 'Midland Bank', money inside, open doors, 124 mm	£15	£12	£7	£5
No 282 Austin Taxi. Blue, red interior, opening bonnet, boot, taxi sign on top, decals on sides, white square on door, 101 mm	£6	£5	£3	£1
No 281 Pathe News Camera Car. Yellow, man and camera, opening boot and bonnet, roof stand, black decals on side, 108 mm	£10	£7	£3	£2
No 287 Police Accident Unit. Cream, red, opening doors (sliding and opening back doors), driver, police decals on roof and sides, road signs, 122 mm	£10	£9	£5	£3
No 289 Routemaster Bus. Red double decker, driver, conductor, Schweppes decals on top sides, London Transport on lower sides, signs on front and back, 121 mm	£15	£12	£6	£4
No 293 Atlantean (Shell BP Bus). Green, cream, driver, conductor, decals on top sides, BP IS THE KEY TO BETTER MOTORING, Lower decals CORPORATION TRANSPORT, 121 mm, Please note colours vary on this model, red/cream, blue	£40	£30	£10	£5
No 952 Vega Major Luxury Coach. Cream, chocolate or maroon trim, six wheels, World decal on side, battery for lights and bell, 245 mm	£25	£20	£10	£5
No 987 ABC TV Mobile Control Room. Light blue, grey, man, camera, red, white trim, decals on side, ABC television, 151 mm	£25	£20	£6	£3
No 978 Refuse Wagon. Green cab, white body or silver, two bins, black tyres, red hubs, tips straight up, slants, decals on side, 162 mm	£20	£15	£5	£2
No 988 ABC Transmitter Van. Light blue, grey underbody, red, white trim, transpeaker on roof, 113 mm	£30	£25	£6	£3
No 257 Fire Chief's Car. Red, light on roof, 'Fire Chief' decal on door, 102 mm	£10	£8	£5	£3
No 259 Fire Engine. Red, ladder, hose, 117 mm	£15	£12	£6	£4
No 276 Airport Fire Tender with Flashing Light. Red, bell, hose, opening doors, 117 mm	£20	£18	£10	£5
No 955 Fire Engine with Extending Ladder. Red, bell two-way adjustment on back roof, 140 mm	£20	£18	£10	£5
No 596 Turntable Fire Escape. Red, moving ladder, 197 mm	£15	£12	£8	£5
No 324 Hayrake. Red, silver rake, yellow wheels, 76 mm	£3	£2	£1.50	£1
No 300 Massey Ferguson Tractor. Red, driver, yellow wheels, Massey Ferguson decals, 89 mm	£10	£7.50	£5	£3
No 319 Weeks Tipping Farm Trailer. Red, yellow, black tyres, red hubs, 105 mm	£5	£4	£2	£1

DINKY — SELECTED SPECIALS

	Mint boxed	Mint unbxd	Good cond	Fair cond
No 320 Halesowen Harvest Trailer. Red, yellow wheels, yellow gates, 133 mm	£5	£4	£2	£1
No 321 Massey Harris Manure Spreader. Red, yellow wheels, silver forks, silver cutters, 121 mm	£4	£3	£2	£1
No 305 David Brown Tractor. Silvery grey, red funnel, 83 mm	£7	£6	£4	£2
No 325 David Brown Tractor with Disc Harrow. Grey, red trim, 152 mm	£10	£8	£5	£3
No 322 Disc Harrow. Grey, red, 79 mm	£2	£1.50	£1	50p
No 341 Land Rover Trailer. Red, orange, 79 mm	£5	£4	£3	£2
No 340 Land Rover. Orange or red, driver, green interior, 92 mm	£10	£7	£4	£2
No 342 Austin Mini Moke. Green, silver trim, off white roof, opening bonnet and lift off hood, 73 mm	£5	£4	£2	£1
No 448 Chevrolet Pickup and Trailers. Light blue, white, red trailers, one open, one boxed, 111 mm	£10	£8	£5	£3
No 449 Chevrolet El Camio Pick Up Truck. Blue, white, 111 mm	£6	£5	£3	£1
No 601 Austin Para Moke. Parachute lift off hood, 76 mm	£5	£4	£2	£1
No 603 Army Private. Seated	50p each			
No 674 Austin Champ. Dark green, driver, spare wheel on back, 70 mm	£5	£4	£3	£1
No 651 Centurian Tank. Green, Army decals on front, rear, 146 mm	£10	£6	£4	£3
No 665 Honest John Missile Launcher. Army green, decals, missile that fires, 188 mm	£10	£8	£5	£3
No 697 25 Pounder Field Gun Set. Green, 79 mm, 89 mm	£15	£12	£7	£5
No 670 Armoured Car. Green, decals on corners, turning turret, 73 mm	£6	£5	£3	£2
No 279 Aveling Barford Diesel Roller. Orange, green, tan driver, 116 mm	£10	£8	£5	£3
No 435 Bedford T K Tipper. Yellow, silver, silver trim, red wheels, tipper, 121 mm	£10	£9	£6	£5
No 436 Atlas Copco. Mustard, compressor lorry, Copco decals, 89mm	£6	£5	£3	£1
No 437 Muirhill 2 W L Loader. Red, black, driver, grey shovel, 121 mm	£7	£6	£5	£3
No 925 Leyland Dump Truck. Grey cab, red body, 12 wheels, tilt cab, 192 mm	£15	£14	£10	£8
No 959 Foden Dump Truck. Red, grey, driver, bulldozer, tips, 165 mm	£15	£14	£10	£8
No 960 Albion Lorry with Concrete Mixer. Red, spare wheel on side, blue, yellow mixer, 130 mm	£10	£7	£5	£4
No 964 Elevator Loader. Blue, yellow, 156 mm	£6	£5	£3	£2
No 965 Euclid Dump Truck. Yellow, 'Stone Ore Earth' decals on side, 'Euclid' on door, 143 mm	£10	£8	£5	£3
No 970 Jones Fleetmaster Crane. Red, dark tan, 'Jones Fleetmaster' decals on side, 'Jones' black decal on top, front, 174 mm, Cantilever	£20	£15	£5	£4
No 972 Coles 20 Ton Crane. Red wagon, orange crane, mounted on lorry, 245 mm	£25	£20	£8	£5
No 975 Rushton Bucyrus Excavator. Yellow, green, black driver, 190 mm	£15	£12	£8	£5
No 402 Bedford Coca Cola Truck. Red; silver grey or white trimming, detachable load; 'Coca Cola' decal along white board on topside; 121 mm	£20	£15	£5	£3
No 407 Ford Transit Van. Light blue, silver trimming; opening doors, 'Kenwood' decal on side, 122 mm	£15	£12	£8	£5
No 425 Bedford Coal Wagon. Red, silver, T K model six coal sacks full of coal (scales), 'Approved coal merchant' decals on side of door, 121 mm	£25	£20	£10	£7
No 434 Bedford T K Crash Truck. Red cab; black roof; black tow hook, grey crane, 'Auto Services' decal on side, 124 mm	£20	£15	£5	£3
No 450 Bedford Box Van. T K Model, Green, white panels, sliding doors at sides and back, 'Castrol' decals on front, sides in red letters, 143 mm	£25	£20	£5	£3
No 914 AEC Articulated Lorry Red cab, grey wagon, green detachable cover, 'British Road Services' decal on cover, 210 mm	£25	£20	£7	£5

	Mint boxed	Mint unbxd	Good cond	Fair cond
No 936 Leyland 8 Wheeled Chassis. Red cab, silver body, driver, 3 weight blocks, '5 tons' decals on side of each block, 197 mm	£30	£25	£8	£5
No 944 Shell B P Fuel Tanker. Silver grey or white with yellow body, cab, 'Shell BP' decals on side, 194 mm	£40	£30	£10	£5
No 945 AEC Fuel Tanker. Silver, red trimming, or white, red trimming, 'Esso' decals on sides, 266 mm	£40	£30	£10	£8

Rolls Royce Silver Shadow

Lancing Bagnal tractor

Dublo-dinky toys

	Mint boxed	Mint unbxd	Good cond	Fair cond
No 062 Singer Roadster. Yellow, red interior, 51 mm	£10	£8	£3	£2
No 063 Commer Van. Blue, silver trim, 54 mm	£10	£8	£5	£3
No 064 Austin Lorry. Green; black wheels, 64 mm	£10	£8	£5	£3
No 067 Austin Taxi. Blue, cream, driver, 57 mm	£10	£8	£5	£3
No 068 Royal Mail Van. Red, windows, 'Royal Mail' decals on sides, 48 mm	£50	£40	£15	£10
No 069 Massey Harris Ferguson Tractor. Light blue, grey wheels, 28 mm	£10	£8	£5	£3
No 070 AEC Mercury Tanker. Green cab, red body, Shell B P decals on sides, 92 mm	£10	£8	£5	£3
No 071 Volkswagen Delivery Van. Golden yellow, windows, 'Hornby Dublo' decals on sides, 54 mm	£10	£8	£5	£3
No 072 Bedford Articulated Flat Truck. Yellow cab, red trailer, windows, 117 mm	£10	£9	£6	£5
No 073 Land Rover with Horse Trailer. Green land-rover, orange horsebox, horse, windows, 117 mm	£10	£8	£5	£3
No 076 Lansing Bagnall Tractor. Maroon; blue driver, trailer, 104 mm	£6	£5	£3	£2
No 066 Bedford Flat Truck. Grey, 64 mm	£10	£5	£3	£2

Dinky specials of the fifties

Model	Mint boxed	Mint unbxd	Good cond	Fair cond
No 105 Triumph TR2 Sports. Yellow, driver, Touring finish, black wheels, 2/8d, 85 mm	£10	£8	£5	£3
No 132 Packard Convertible. Green, red interior, black driver, silver trim, 3/10d, 115 mm	£10.	£8	£5	£3
No 150 Rolls Royce Silver Wraith. Grey, off white, windows, four wheeled suspension, 5/4d, 121 mm	£15	£12	£10	£7
No 157 Jaguar XK 120. Red coupe, matching trim, 2/5d, 97 mm, very special model	£15	£12	£9	£5
No 160 Austin A30 Saloon. Orange or deep salmon, silver trim, grey wheels, 2/3d, 76 mm	£7	£5	£3	£2
No 165 Humber Hawk. Black, green trim around centre, windows, four wheeled suspension, 3/9d, 102 mm	£10	£7	£5	£3
No 166 Sunbeam Rapier. Two-tone blue, windows, 2/11d, 82 mm	£10	£8	£5	£3
No 167 A C ACECA Coupe. Grey body, red roof, windows, 2/11d; 89 mm	£10	£8	£5	£3
No 168 Singer Gazelle. Two-tone green bottom, grey top, windows, 2/11d, 79 mm, rare model	£15	£12	£7	£5
No 169 Studebaker Golden Hawk. Mustard, green trim, red fins, windows, 3/5d, 108 mm	£10	£5	£3	£2
No 173 Nash Rambler. Blue, red trim, windows, 3/5d, 115 mm	£10	£5	£4	£2
No 174 Hudson Hornet. Red, yellow roof, yellow trim, grey wheels, 3/5d, 111 mm	£10	£7	£5	£3
No 175 Hillman Minx. Yellow, green top, windows, silver grey wheels, 2/11d, 79 mm	£8	£5	£3	£2
No 176 Austin 105 Saloon. Cream, blue roof; blue trim, windows, 2/11d, 102 mm	£7	£5	£3	£2
No 178 Plymouth Plaza. Light blue, dark blue roof, dark blue trim, windows, 3/5d, 105 mm	£9	£8	£5	£3
No 179 Studebaker President. Yellow, blue fins, windows, 3/5d, 105 mm	£7	£6	£4	£2
No 180 Packard Clipper. Orange, silver grey roof, black trim, windows, 3/5d, 111 mm	£10	£9	£5	£3
No 181A Special Volkswagen. Deep purple, 2/5d, 85 mm, rare model	£15	£14	£10	£5
No 182 Porsche 356A. Light blue, silver trim, coupe with windows, 2/10d, 89 mm	£10	£9	£7	£5
No 183 Fiat 600 Saloon. Red, grey wheels, 2/3d, 54 mm	£15	£14	£10	£7
No 187 Volkswagen Karmann. Red, black roof; Ghia coupe, windows, four wheel suspension, 3/6d; 96 mm	£15	£14	£10	£5
No 188 Four Berth Caravan. Green, cream, tow hook, windows, interior fittings, 2/6d, 130 mm	£15	£14	£10	£5
No 189 Triumph Herald. Green, silver trim, windows, four wheel suspension, 3/3d, 79 mm	£15	£14	£10	£5
No 190 Caravan. Orange, cream, 3/8d, 121 mm, model suitable for attachment to most Dinky cars	£10	£8	£6	£4

DINKY SPECIALS OF THE FIFTIES 105

	Mint boxed	Mint unbxd	Good cond	Fair cond
No 191 Dodge Royal Sedan. Primrose, orange trim, grey wheels, 3/5d, 111 mm	£15	£12	£10	£7
No 192 De Soto Fireflite. Light blue, orange roof, orange, yellow trim, 3/5d, 111 mm	£15	£14	£10	£7
No 195 Jaguar 34 Litre Saloon. Primrose, silver trim, four wheel suspension, steering wheel, 3/5d, 79 mm	£15	£12	£10	£7
No 230 Talbot Lago Racing Car. Blue, white driver, silver trim, 2/5d, 102 mm	£10	£9	£5	£3
No 231 Maserati Racing Car. Red, white driver, yellow, silver trim, 2/5d, 91 mm	£10	£9	£5	£3
No 232 Alfa Romeo Racing Car. Red, white driver, silver trim, number 8, 2/5d, 102 mm	£10	£9	£5	£3
No 233 Cooper Bristol Racing Car. Green, white driver, grey wheels, number 6, 2/5d, 89 mm	£10	£9	£5	£3
No 234 Ferrari Racing Car. Dark blue, yellow nose, white driver, grey wheels, 2/5d, 102 mm	£10	£9	£5	£3
No 237 Mercedes Benz Racing Car. White; grey driver, silver trim, number 30, 2/11d, 97 mm	£15	£14	£10	£7
No 238 Jaguar D Type Racing Car. Light green, white driver, 2/11d, 79 mm	£10	£9	£5	£3
No 239 Vanwall Racing Car. Green, white driver, vanwall decals no 35, 2/11d, 96 mm	£10	£9	£5	£3
No 250 Streamlined Fire Engine. Red, ladder, bell, 3/2d, 102 mm	£15	£14	£10	£7
No 251 Aveling Barford Diesel Roller. Green; orange driver, red, grey wheels, 4/10d, 111 mm	£15	£14	£10	£7
No 252 Refuse Wagon. Orange, green sliding doors, Bedford chassis, tipping body, 6/4d, 108 mm	£20	£18	£12	£7
No 253 Daimler Ambulance. Silver grey, red cross decals on side, red wheels, 2/11d, 96 mm	£20	£18	£12	£7
No 254 Austin Taxi. Black or yellow, driver, 3/5d, 79 mm	£20	£15	£12	£10
No 255 Mersey Tunnel Police Van. Red, silver trim, mersey tunnel decals, 2/10d, 72 mm	£15	£14	£10	£5
No 256 Police Patrol Car. Black, two policemen, windows, seat, four wheel suspension, 3/11d, 102 mm	£25	£20	£10	£5
No 257 Fire Chiefs Car. Red, silver trim, fire chief decals, roof light, 3/5d, 102 mm	£25	£20	£10	£5
No 258 USA Police Car. Black, white trim, windows, seats, four wheel suspension, steering wheel, police decals, 2/11d, 111 mm	£10	£9	£7	£5
No 260 Royal Mail Van. Red, black roof, Post Office Royal Mail decals, 2/10d, 79 mm; very rare	£25	£20	£15	£10
No 261 Telephone Service Van. Green, black roof, white ladder, correct post office telephone decals, 2/10d, 72 mm, very rare	£25	£20	£15	£10
No 265 Plymouth U S A Taxi. Orange, red roof, windows, seats, 4 wheel suspension, steering wheel, taxi sign decals, 4/-d, 105 mm	£15	£12	£10	£7
No 270 AA Motor Cycle Patrol. Yellow, tan driver, black cycle, wheels, correct AA decals, 2/2d, 28 mm	£15	£12	£10	£7
No 282 Duple Roadmaster Coach. Red, silver trim, 3/5d, 121 mm, very rare	£40	£30	£15	£10
No 283 BOAC Coach. Deep purple, white roof, silver trim, light blue wheels, correct BOAC decals, 4/2d, 121 mm, very rare	£25	£20	£15	£10
No 290 Double Decker Bus. Cream top, red bottom, Dunlop decals, 4/2d, 102 mm	£30	£25	£15	£10
No 291 London Bus. Red, Silver trim; Exide decals, 4/2d, 102 mm	£25	£20	£15	£10
No 295 Atlas Bus. Windows, seating, steering wheel, four wheel suspension, Standard Atlas Kenebrake, 3/9d, 85 mm	£15	£12	£8	£5
No 300 Massey Harris Tractor. Red, driver, yellow trim, yellow wheels grey tyres, correct decals, 4/2d, 89 mm	£15	£12	£8	£5
No 301 Field Marshall Tractor. Orange; tan driver; silver trim, adjustable steering, green wheels grey tyres, 4/2, 76 mm	£20	£15	£10	£5
No 340 Land Rover. Orange, tan driver, red wheels, 4/5d, 91 mm	£15	£12	£10	£7
No 341 Land Rover Trailer. Orange, red wheels grey tyres, 2/-d, 79 mm	£6	£5	£3	£2

DINKY SPECIALS OF THE FIFTIES

	Mint boxed	Mint unbxd	Good cond	Fair cond
No 343 Farm Produce Wagon. Green body, yellow top, silver trim, yellow wheels, 3/8d, 108 mm	£15	£12	£10	£8
No 344 Estate Car. Mustard, brown, silver trim, panelled, 2/11d, 105 mm	£15	£10	£8	£5
No 405 Universal Jeep. Green, black trim, 3/2d, 82 mm	£15	£10	£8	£5
No 408 Big Bedford Lorry. Maroon, mustard, silver trim, yellow wheels, grey tyres, 5/3d, 146 mm	£20	£15	£10	£7
No 409 Bedford Artic. Lorry. Orange, black trim, red wheels, 6/7d, 166 mm	£15	£12	£10	£7
No 410 Bedford End Tipper. Red chassis, cream body, black trim, tipping with hinged tailboard, 5/6d, 97 mm	£15	£10	£8	£5
No 413 Austin Covered Wagon. Red, grey cover, black trim, grey wheels, 3/5d, 105 mm	£20	£10	£8	£5
No 414 Rear Tipping Wagon. Red chassis, green body, silver trim, green wheels, tipping body with hinged tail board, 3/5d, 102 mm	£20	£15	£10	£5
No 420 Leyland Forward Control Lorry. Green, silver trim, red wheels, black tyres, 2/5d, 108 mm	£20	£12	£10	£7
No 430 Breakdown Lorry. Mustard chassis; green body, red wheels, grey hook, Dinky Service decals on sides, 5/9d, 123 mm	£15	£12	£10	£5
No 431 Guy Warrior 4 Ton Lorry. Mustard chassis, green body, silver trim; green wheels, 5/9d, 134 mm	£20	£15	£10	£7
No 981 Horse Box. Chocolate or maroon, side, rear ramps hinged, British Railway decals, 12/9d, 174 mm	£18	£15	£10	£7
No 982 Mighty Antar Low Loader. Red chassis, grey body, brown prop, silver grey wheels, 14/11d, 306 mm	£20	£15	£12	£9
No 794 Pullmore Car Transporter. Blue cab, light blue trailer, silver trim, darker blue wheels, black tyres, Dinky Toy Delivery Service decals, 232 mm	£20	£15	£10	£9
No 108 M.G. Midget Sports. Salmon pink, green, driver, no 24, 3/9d, 82 mm	£10	£8	£5	£3
No 133 Cunningham C-5R. Road Racer. White, black stripes; driver, 4/3d, 102 mm, colours can vary	£15	£10	£7	£5
No 421 Hindle Smart Electric Artic. Maroon correct decals, 3/11d, 136 mm	£20	£15	£10	£7
No 902 Foden Flat Truck. Red cab, green body, 7/3d; 187 mm, colours can vary	£25	£15	£10	£8
No 415 Mechanical Horse & Tipper Wagon. Maroon, British Rail decals, 3/1d, 136 mm, colours may vary	£20	£15	£12	£10
No 942 Foden 14 Ton Tanker. Regent decals, correct co, colours, 9/6d, 187 mm, very rare	£40	£35	£15	£12
No 901 Foden Diesel 8 Wheel Wagon. Red or green cab, grey or green body, 7/9d, 187 mm, colours vary, very rare	£40	£35	£18	£15
No 973 Goods Yard Crane. Mustard, orange, yellow, red, 11/-d, 255 mm	£15	£12	£10	£7
No 798 Express Passenger Train. Mallard blue, green, yellow trim, Flying Scotsman, 7/6d, 382 mm	£15	£12	£10	£7
No 441 Tanker Castrol. Decals, company correct colours, 2/10d, 111 mm, very rare	£40	£30	£15	£10
No 450 Trojan 15 Cwt Van. Silver trim, Esso decals, colours, 2/6d, 85 mm, very rare	£30	£25	£15	£10
No 451 Trojan 15 Cwt Van. Dunlop decals, company colours, 2/6d, 85 mm, very rare	£25	£20	£15	£10
No 452 Trojan 15 Cwt Van. Yellow trim, company decals in gold, Chivers Jellies, Green, 2/6d, 85 mm, very rare	£25	£20	£15	£10
No 481 Bedford 10 Cwt Van Ovaltine. Decals, in company colours, 2/9d, 102 mm, very rare	£40	£30	£15	£10
No 918 Guy Van. Dark blue, red, white trim, Ever Ready decals, 7/6d, 127 mm, very rare	£40	£35	£12	£10
No 432 Guy Warrior Flat Truck. Green chassis, red body, red wheels, black tyres, 5/3d, 134 mm	£40	£35	£9	£7
No 440 Tanker Mobilgas. Red, silver trim, Mobilgas decals, 2/10d, 123 mm	£40	£35	£10	£5
No 455 Trojan 15 Cwt Van. Red, silver trim, Brooke Bond decals on side, 2/5d; 85 mm, very rare	£25	£20	£10	£7
No 471 Austin Van Nestles. Red, silver, yellow trim, yellow wheels, Nestles decals on side, 2/5d, 89 mm,	£25	£20	£10	£7
No 472 Austin Van. Green, silver trim, yellow wheels,				

	Mint boxed	Mint unbxd	Good cond	Fair cond
black tyres; Raleigh decals in yellow, 2/5d, 89 mm	£25	£20	£10	£7
No 905 Foden Flat Truck. Red, silver trim, black tyres; silver grey flat, chains, 9/6d, 184 mm	£40	£30	£15	£10
No 930 Bedford Pallet Jekta Van. Orange, yellow, yellow wheels, Meccano, Dinky Toy decals, 14/3d, 168 mm	£40	£25	£10	£7
No 934 Leyland Octopus Wagon. Yellow chassis, green body, green trim, red wheels, black tyres, 14/3d, 184 mm	£40	£30	£10	£7
No 943 Leyland Octopus Tanker. Red, white trim, ladder top, fittings, Esso decals, 10/6d, 190 mm	£50	£30	£10	£8
No 961 Blaw Knox Bulldozer. Yellow, grey, black, 14/6d, model runs on creeper tracks, 142 mm	£20	£15	£10	£7
No 962 Muir Hill Dump Truck. Yellow, tan driver, silver grey trim, red wheels, 8/3d, 105 mm	£15	£12	£9	£6
No 971 Coles Mobile Crane. Golden yellow, black, yellow wheels, grey tyres, black hook, 11/-d, 168 mm, jib 142 mm	£40	£30	£10	£7
No 972 Coles 20 Ton Mounted Crane. Salmon, orange, silver trim, grey tyres, black hook, 16/-d, 156 mm x 244	£40	£30	£10	£8
No 967 B B C T V Control Room. Green, white, yellow trim, windows, decals, 8/3d, 130 mm	£40	£30	£10	£7
No 968 B B C T V Roving Eye. Green; yellow, white trim; windows; camera man, decals, 8/3d, 105 mm	£40	£30	£15	£10
No 969 B.B C T V Extending Ladder & Mast Vehicle. Green, 13/6d, 166 mm, These are exact models of mobile camera units of B B C Finished in dark green with B B C coat of arms	£50	£40	£15	£10
No 984 Car Carrier & Trailer 985. Red, grey roof, car, Dinky decals, £1.2s and 16/-d, 232 mm, carrier 229 mm	£60	£50	£15	£10

Dinky specials - the British Army

	Mint boxed	Mint unbxd	Good cond	Fair cond
No 603 Army Personnel. 6d each, seated or otherwise	£1	75p	50p	25p
No 641 Army 1 Ton Cargo Truck. 3/9d, standard colours, 79 mm	£6	£5	£3	£2
No 621 3 Ton Army Wagon. 5/3d, 115 mm	£15	£10	£8	£5
No 622 10 Ton Army Truck. 7/-d, 178 mm	£20	£15	£10	£5
No 623 Army Covered Wagon. 4/2d, 105 mm	£15	£10	£5	£3
No 643 Army Water Tanker. 4/2d, 89 mm	£15	£10	£5	£3
No 626 Military Ambulance. 6/4d, red cross decals, 111 mm	£15	£10	£7	£5
No 677 Armoured Command Vehicle. 5/4d, 134 mm	£15	£10	£8	£5
No 676 Army Personnel Carrier. 4/6d, 127 mm	£15	£10	£7	£5
No 699 Military Gift Set. 17/6d, four models, very rare	£30	£25	£10	£9
No 670 Armoured Car. 2/10d, 72 mm	£10	£8	£5	£3
No 675 Small Armoured Personnel Carrier. 3/8d, 82 mm	£10	£7	£5	£3
No 651 Centurian Tank. 8/6d, revolving gun turret, 146 mm	£20	£15	£10	£5
No 661 Recovery Tractor. 8/9d, towing unserviceable military lorries, 134 mm	£20	£15	£10	£5
No 689 Medium Artillary Tractor. 8/9d; an excellent model of a 6 wheel drive, 140 mm	£20	£15	£10	£5
No 660 Tank Transporter. 13/11d, for transporting centurian tank, 338 mm	£25	£20	£10	£5

108 DINKY SPECIALS — THE BRITISH ARMY

	Mint boxed	Mint unbxd	Good cond	Fair cond
No 666 Missile Erector Vehicle. £1.8.9d with corporal missile and launching platform, vehicle 232 mm, missile 229 mm	£25	£20	£15	£10
No 667 Missile Servicing Platform. 13/6d, 198 mm	£15	£10	£7	£5
No 977 Servicing Platform Vehicle. 13/6d, 198 mm	£15	£10	£7	£5
No 697 25 Pounder Field Gun Set. 7/3d, 213 mm	£20	£15	£10	£5
No 693 7.2 Howitzer. 3/5d, 130 mm	£8	£7	£5	£3
No 673 & 674 Scout Car & Austin Champ. 3/2d each, 66 mm and 70 mm	£6	£5	£3	£2

Hansom cab

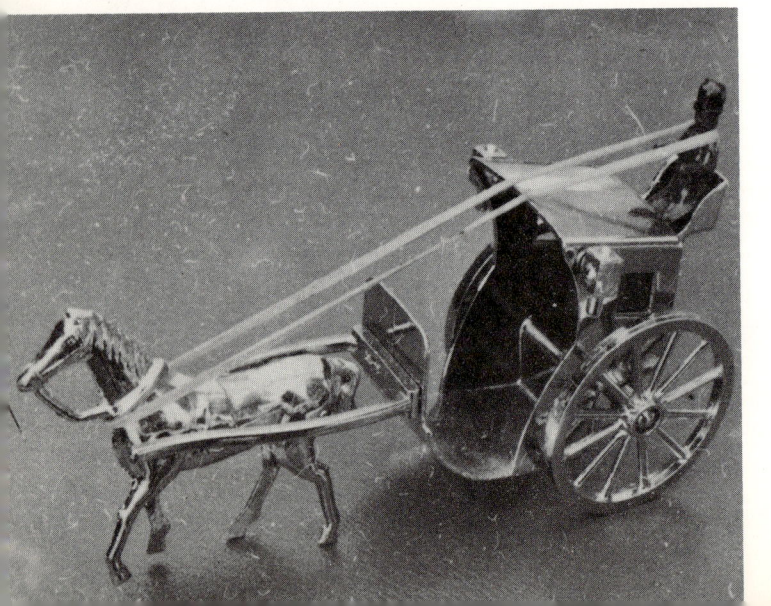

	Mint boxed	Mint unbxd	Good cond	Fair cond
No 705 Vickers Viking Airliner. Silver, blue trim, black letters, red props, 2/-d, wing span 140 mm	£10	£9	£7	£5
No 708 Vickers Viscount 1800 BEA. Silver, red trim; BEA decals, red props, 4/8d, wing span 148 mm	£15	£10	£7	£5
No 715 Bristol 173 Helicopter. Blue, red trim, black letters, red props, 2/8d, 127 mm	£15	£10	£7	£5
No 716 Westland Sikorsky Helicopter. Red, silver trim, gold letters, red props, 2/5d, 89 mm	£10	£8	£5	£3
No 732 Meteor Twin Jet Fighter. Silver trim, R A F decals, 1/-d, wing span 66 mm	£10	£8	£5	£3
No 733 Shooting Star Jet Fighter. Silver, blue trim, white stars, 1/-d, wing span 60 mm	£10	£8	£5	£3
No 734 Supermarine Swift Fighter. Correct milt, decals, colours, 1/9d, wing span 51 mm	£10	£8	£5	£3
No 735 Gloster Javelin Delta Wing Fighter. Correct RAF colours, decals, 2/5d, wing span 82 mm	£10	£8	£5	£3
No 736 Hawker Hunter Fighter. Correct RAF decals, colours, 1/9d, 54 mm	£10	£8	£5	£3
No 737 Lightning P IB Fighter. Silver; black nose; RAF decals, correct ensign, 2/-d, 54 mm	£15	£10	£8	£5
No 998 Bristol Brittania. Silver, red strips, blue decals 'Canadian Pacific', 9/3d, wing span 225 mm	£15	£12	£9	£5
No 999 D H Comet Airliner. Silver, Blue trim, black letters, B O A C decals, 6/4d, wing span 184 mm	£15	£12	£10	£7
No 700 Seaplane. Authentic colours, company markings, 1/5d, 102 mm	£15	£10	£7	£5

Dinky specials - winged wonders

London bus

	Mint boxed	Mint unbxd	Good cond	Fair cond
No 704 Avro York Airliner. White, red or blue trim, black letters, co. colours, 2/11d, 102 mm	£15	£10	£7	£5
No 731 Twin Engined Fighter. Correct RAF colours can be bought in sets of three, 8d, 51 mm	£6	£5	£3	£1
No 730 Tempest Fighter. Correct RAF decals, colours, 9d, 51 mm, a small well sought after plane	£10	£6	£4	£2
No 204 Volkswagon Pick Up Truck. Blue body, Express delivery written on side, 96 mm, 3oz	£10	£8	£5	£3
No 214 Mobile Salvage Crane. Red, yellow, blue, 108 mm, 3oz	£15	£10	£5	£3
No 220 Cattle Truck. Red, brown, 96 mm, 4¾oz	£15	£10	£5	£3
No 228 Coca Cola Van. twelve crates, 140 mm, 7¾oz	£15	£13	£10	£5
No 232 Cable Drum Transporter. Red, brown, yellow, 172 mm, 1½oz	£20	£15	£10	£7
No 240 British Railways Articulated Van. Red, cream, green, canvas top, 153 mm, 4½oz	£15	£12	£7	£5
No 244 Budgie Breakdown Truck. Blue, red crane 140 mm, 5oz	£12	£10	£6	£4
No 246 Police Patrol Car. Blue, siren on top, 115 mm, 4oz	£12	£10	£7	£5
No 254 Merryweather Fire Engine Turntable. Red, windows, ladder fire escape, 108 mm, 4½oz	£20	£15	£10	£5
No 256 Esso Aircraft Refuelling Tanker. Pluto with windows, red with Esso on sides, 140 mm, 6½oz	£20	£18	£10	£6

	Mint boxed	Mint unbxd	Good cond	Fair cond
No 260 Rushton Bucyrus Excavator. Red, yellow, green, 184 mm, 8½oz	£10	£8	£5	£3
No 264 Racing Sidecar Unit. Red black riders, 108 mm, 3½oz	£10	£7	£5	£3
No 266 Express Delivery Sidecar Outfit. Red, blue, brown; black rider, 108 mm. 5oz	£15	£12	£7	£5
No 268 AA Patrol Landrover with Windows. Yellow, black; correct lettering as real model, 96 mm, 5oz	£20	£15	£10	£5
No 270 Esso Tanker with Windows. Red, ladders, correct lettering, 134 mm, 4¼oz, 00 scale	£25	£20	£12	£6
No 272 Supercar. 127 mm	£10	£8	£5	£3
No 274 Refuse Truck. Yellow, silver, windows; 121 mm, 6oz	£25	£20	£7	£5
No 276 Bedford LWB Tipper. Red, yellow, windows, 127 mm, 5oz	£25	£20	£10	£6
Budgie Special London Bus. Red; decals on side, front, rear, various routes in London area, Made for H Seener & Co	£10	£7	£2	£1
Budgie Special Hansom Cab. Silver, Gold, Red driver, black cab, silver horse, made for H Seener & Co	£5	£4	£2	£1
Budgie Special Red Railway Engine. Red, adapted to fit electric motor; 00 scale, made for H Seener & Co	£5	£4	£3	£2

DINKY SPECIALS — WINGED WONDERS

Corgi, another household name in the world of diecast models, produced many attractive vehicles.

	Mint boxed	Mint unbxd	Good cond	Fair cond
No 238 Jaguar Mk 10 Saloon. Blue, silver trim, opening bonnet, boot, luggage, 6/11d, 108 mm. model of models; thoroughbred of miniatures, Corgi model loaded with features, graceful power of original, long flowing lines, entirely new 'first' in boot — Luggage you take out, open, press front of bonnet down, bonnet opens to reveal detailed engine, note twin headlights, test glidomatic spring suspension, seats, steering wheel, windows, no finer models in world produced such as these	£10	£8	£5	£3
No 437 Superior Ambulance on Cadillac Chassis. Red, dark yellow or cream roof, silver trim, red flashing light, decals on sides, 8/9d, 115 mm, This Corgi model has another 'first'. Five battery-operated flashing lights! But has only one bulb illuminating all the lights. The battery lasts for ages. Changing battery or bulb takes only a few seconds. All you have to do is to unclip the power pack, lift out bulb or battery, put new one in, then clip power pack back into place. Once in it stays there! The model has a cutout switch so you can stop the lights flashing, when you like, to avoid battery wastage. No other model has such a power pack system. Model also has seats, windows, steering wheel and the fantastic Glidomatic spring suspension. This has been a best seller of all time.	£15	£12	£10	£7

	Mint boxed	Mint unbxd	Good cond	Fair cond
No 224 Bentley Continental. Two tone green, silver trim, plated radiator, bumpers, detachable spare wheel, steering wheel, seats, jewelled headlights, ruby rear lights, 7/8d, 108 mm,	£15	£12	£9	£5
No 218 Aston Martin D.B.4. Deep yellow, silver trim, seats, steering wheel, opening bonnet, diecast engine, glidomatic spring suspension, 4/7d, 96 mm	£15	£10	£7	£5
No 231 Triumph Herald Coupe. Brown, silver; green trim, seats, steering wheel, glidomatic spring suspension, opening bonnet, fully detailed engine, 5/3d, 89 mm	£15	£10	£7	£5
No 225 Austin Seven. Red, silver trim, seats, steering wheel, glidomatic suspension, 3/5d, 72 mm, finest finish any model could have, five star award winner	£12	£10	£6	£4
No 234 Ford Consul Classic. Silver pink, red roof, opening bonnet, detailed engine, suspension, 4/9d, 96 mm	£10	£7	£5	£3
No 233 Trojan Heinkel Bubble Car. Red, silver trim; seats, steering wheel, glidomatic spring suspension, 3/-d, 64 mm	£7	£5	£3	£2
No 424 Ford Zephyr Estate Car. Light blue, dark blue bonnet, white or silver trim, seats, steering wheel, glido. suspension, moulded luggage, 4/-d, 97 mm	£10	£7	£5	£3
No 208 S Jaguar 2-4 Litre. Mustard or primrose seats, steering wheel, glido. suspension, 3/9d, 96 mm	£10	£7	£5	£3

Corgi

Item	Mint boxed	Mint unbxd	Good cond	Fair cond
No 226 Morris Mini Minor. Blue, silver trim, seats, steering wheel, glido. suspension, 3/5d, 72 mm	£10	£7	£5	£3
No 439 Chevrolet Fire Chief Car. Red; silver trim; seats, steering wheel, glidomatic suspension, decals on side, front, two figures, aerial, 108 mm	£15	£10	£6	£4
No 430 Bermuda Taxi. White, seats, steering wheel, glidomatic spring suspension, driver, detachable sun canopy blue, red, taxi decals, 5/3d, 102 mm	£15	£10	£6	£4
No 237 Oldsmobile Sheriff Car. White, black roof, bonnet, seats, steering wheel, glidomatic spring suspension, Sheriff decals, 4/11d, 108 mm	£10	£8	£5	£3
No 219 Plymouth Sports Suburban. Mustard yellow; seats, steering wheel, 3/10d, 105 mm	£10	£8	£5	£3
No 223 Chevrolet State Patrol Car. Black, seats, steering wheel, glidomatic spring suspension, aerial, police decals, 4/7d, 108 mm	£10	£8	£5	£3
No 215S. Ford Thunderbird Open Sports. Red, driver, seats, steering wheel, glidomatic spring suspension, 4/4d, 102 mm	£10	£8	£5	£3
No 214S. Ford Thunderbird. Black, red roof, seats, steering wheel, glidomatic spring suspension, 4/-d, 102 mm	£10	£7	£5	£3
No 211S. S. Studebaker Golden Hawk. Golden, white trim, seats, steering wheel, glido. suspension, 4/7d, 105 mm	£10	£8	£6	£4
No 220 Chevrolet Impala. Pink, white, silver trim, white wheels, glido. suspension, 4/1d, 108 mm	£10	£8	£6	£4

Item	Mint boxed	Mint unbxd	Good cond	Fair cond
No 229 Chevrolet Corvair. Bright blue, silver trim, seats, steering wheel, glido. suspension, venetian blind, opening cover at rear showing detailed engine, 4/9d, 97 mm	£10	£7	£5	£3
No 221 Chevrolet Yellow Cab. Yellow; silver trim; seats, steering wheel, glido suspension, aerial; taxi sign, 4/9d, 108 mm	£10	£8	£6	£4
No 235 Oldsmobile Super 88. Blue, silver trim, seats, steering wheel, glido. suspension, 4/11d, 108 mm	£10	£8	£6	£4
No 232 Fiat 2100. Yellow, red roof; seats; steering wheel, glido. suspension, jewelled headlights, venetian blind, 4/7d, 96 mm	£10	£8	£5	£3
No 210S. Citroen D S 19 Dark red, seats, steering wheel, glido. suspension, 3/11d, 97 mm	£15	£10	£7	£5
No 230 Mercedes Benz 220 S E Coupe. Rich wine, silver trim, self centre steering, glido. suspension, opening boot, spare wheel, 5/6d, 102 mm	£10	£8	£5	£3
No 222 Renault Floride. Green, silver trim, seats, steering wheel, glido. suspension, 3/7d, 91 mm	£10	£8	£6	£4
No 228 Volvo P 1800. Metalic bronze; silver trim; seats, steering wheel, glido. suspension, jewelled headlights, 4/5d, 96 mm	£10	£8	£6	£4
No 217 Fiat 1800. Blue, silver trim, seats, steering wheel, glido. suspension, 3/7d, 96 mm	£10	£7	£5	£3
No 418 Austin Taxi. Black or blue, silver trim, seats, steering wheel, glido. suspension, 4/-d, 97 mm	£15	£10	£6	£4

CORGI

	Mint boxed	Mint unbxd	Good cond	Fair cond
No 227 Mini Cooper Competition. Blue, silver, spare wheel, jewelled headlights, glido. suspension, 4/3d, 73 mm	£10	£7	£5	£3
No 436 Citroen Safari. ID 19. Mustard; silver trim; opening rear doors, hinged rear seats, remote controlled, moulded interior, steering wheel, glido. spring suspension, Wild Life preservation decal on bonnet, 6/6d, 115 mm	£15	£10	£7	£5
No 309 Aston Martin Comp. Model. Light blue; white roof, seats, steering wheel, suspension, jewelled headlights, opening bonnet revealing detailed engine, 5/6d, 96 mm	£12	£10	£6	£4
No 307 E Type Jaguar. Red, silver trim, detailed interior, glido. suspension, detachable hardtop, 5/3d, 96 mm, one of the best models ever made	£12	£10	£7	£5
No 303 S. Mercedes 300 SL Open. Silver, driver, decals, glido. suspension, 3/3d, 96 mm	£10	£8	£5	£3
No 302 M.G A. Red, blue and silver tint; 2/9d; 89 mm	£10	£8	£5	£3
No 304S. Mercedes Benz 300 SL Hardtop. Silver body, red top, No 7, 4/3d, 96 mm	£10	£7	£5	£3
No 305 Triumph TR3. Green; silver trim, 2/9d, 85 mm	£10	£8	£6	£4
No 154 Ferrari Formula 1. Red, driver, glidomatic spring suspension, detailed rear engine, grand prix, no 36, 3/3d, 91 mm	£7	£5	£3	£1
No 152S. BRM Racing Car. Blue green, driver, Union Jack decal on bonnet, glido suspension, no 17, 4/3d, 91 mm	£10	£8	£6	£4
No 150S. Vanwall Racing Car. Red, driver, silver trim, blue triangle decal no 25, glido. suspension, 4/3d, 91 mm	£10	£8	£6	£4
No 151A. Lotus X1 Racing Car. Yellow, driver, green with red flash in centre; glido. suspension, 4/3d, 82 mm	£10	£7	£5	£3
No 153A. Proteus Campbell Bluebird Record Car. Blue, driver, silver trim, Union Jack, American flags on nose, red flashes at rear, 4/7d, 127 mm	£15	£10	£7	£5
No 16 Gift Set Racing Car, Transporter and Cars. 29/11d, Vanwall, Lotus, BRM racing cars	£25	£15	£10	£8

	Mint boxed	Mint unbxd	Good cond	Fair cond
No 5 Gift Set. 12/9d, BRM, Vanwall, Lotus racing cars in special polyfoam pack	£20	£15	£10	£7
No 1126 Ecurie Ecosse Racing Car Transporter. seats, steering wheel, glido. suspension, sliding door revealing workshop, self centering steering, lowering top deck, 17/6d, 198 mm	£50	£35	£7	£5
No 420 Ford Thames Airborne Caravan. Light green, steering wheel, glido. suspension, opening rear doors, interior furnishing, 6/6d, 96 mm	£15	£10	£8	£6
No 14 Gift Set. Hydraulic tower wagon, single lamp standard, electrician, red, silver trim, yellow bucket, 8/7d; 146 mm lamp standard; 91 mm length of wagon	£15	£12	£9	£6
No 1120 Midland Red Motorway Express Coach. Red, black, silver trim, glido. suspension, steering wheel, super headlights, 34 passenger seats, toilet compartment, moulded chassis details, London Birmingham decals, 8/7d, 140 mm	£50	£30	£15	£10
No 408 Bedford AA Van. Black roof, black trim, blue interior, AA colours, correct decals, 3/7d, 82 mm	£15	£12	£9	£6
No 409 Forward Control Jeep. Blue, red decal on nose, FC/150 with spare wheel, 3/8d, 91 mm	£10	£7	£5	£4
No 4165 R A C Land Rover. Two tone blue or dark blue, seats, steering wheel, R A C decals, glidomatic spring suspension, aerial, 6/6d, 95 mm	£15	£12	£10	£7
No 438 Land Rover. Green, silver trim, light fawn cover, seats, steering wheel, glido. S Suspension, 6/6d, 95 mm	£15	£12	£10	£8
No 417S Land Rover Breakdown Truck. Red; yellow cover on rear, silver trim, light on roof, corgi breakdown service decals on sides, 6/6d, 114 mm	£10	£8	£6	£4
No 419 Ford Zephyr Motorway Patrol. White, seats, steering wheel, glidomatic spring suspension, aerial, moulded equipment details, police decals, 4/8d, 97 mm	£14	£12	£9	£6
No 414 Bedford Military Ambulance. Green, red cross decals on sides, roof, 3/-d, 83 mm, rare model	£20	£15	£10	£5
No 434 Volkswagen Kombi. Light green, white; seats; suspension, detailed interior, 5/10d, 91 mm	£12	£10	£7	£5

CORGI 113

	Mint boxed	Mint unbxd	Good cond	Fair cond
No 460 ERF Neville Cement Tipper. Yellow, silver grey, ERF decals, 5/10d, 95 mm	£15	£12	£10	£8
No 454 Commer Platform Lorry. Primrose, white, fine details, 4/1d, 120 mm	£15	£12	£10	£8
No 457 ERF Double 4G Platform Lorry. Three tone blue, ERF decals on nose, 4/1d, 120 mm	£20	£15	£10	£8
No 458 ERF 64G Earth Dumper. Red chassis; silver trim, yellow dumper, 5/7d, 95 mm	£20	£15	£10	£8
No 433 Volkswagen Delivery Van. Red, white; silver trim; VW decal on nose, 5/11d, 91 mm	£15	£12	£10	£8
No 101 Platform Trailer. Yellow, silver wheels, black tyres, 2/9d, 108 mm	£5	£4	£2	£1
No 452 Commer Dropside Lorry. Deep red chassis, fawn top, silver wheels, 4/3d, 120 mm	£20	£15	£10	£7
No 100 Dropside Trailer. Red or fawn, silver or fawn wheels, tow hook, 2/11d, 108 mm	£5	£4	£2	£1
No 456 ERF 44G. Dropside Lorry. Primrose, white or silver wheels, blue wagon, 4/3d, 120 mm	£15	£12	£10	£7
No 351S R A F Land Rover. Blue, seats, steering wheel, glido s. suspension, correct RAF decals, 5/-d, 95 mm	£15	£10	£7	£5
No 1118 International 6X6 Army Truck. Green, glido spring suspension, 8/7d, 140 mm, rare, much sought after	£20	£15	£10	£7
No 3 Gift Set. RAF Land Rover, Thunderbird guided missile on assembly trolley, RAF colours, decals, 230 mm	£20	£18	£12	£10
No 435 Dairy produce Van. Blue and yellow, sliding door, decals on sides, 5/3d, 102 mm	£15	£12	£10	£7
No 1129 Milk Tanker. Blue and white; detachable driving cab, dairy decals, 9/6d, 191 mm	£15	£12	£10	£7
No 21 Gift Set. Blue and white, ERF Dropside Lorry, platform trailer, two detachable milk churn loads, pack of self adhesive accessories, 9/11d, 236 mm	£25	£15	£12	£10
No 1110 Mobilgas Tanker. Red, detachable driving cab, Mobilgas decals on side, 9/6d, 191 mm, rare	£20	£12	£10	£8

The neat lines of the Corgi Citroen alpine rescue car

CORGI

	Mint boxed	Mint unbxd	Good cond	Fair cond
No 1100 Carriemore Low Loader No 1. Blue, red cab, white drop down loading ramp, haulage winch, 12/8d, 220 mm	£20	£15	£12	£10
No 11 Gift Set. Yellow and dark blue, ERF dropside lorry, platform trailer, cement, plank loads, 9/11d, 226 mm	£20	£15	£12	£10
No 1105 Carriemore Car Transporter. Red cab, blue and green trailer, artic cab, seats, glido s. suspension, driving mirrors, suspension on trailer plus snap action loading ramp, 13/6d, 263 mm	£20	£15	£12	£10
No 1104 Low Loader Machinery Carrier No. 2. Red cab, grey loader, white wheels, black tyres, rear axle, wheels removable for loading heavy machinery, loads hauled with winch, line with hook attached, cab detachable, 12/8d, 220 mm	£20	£15	£12	£10
No 28 Gift Set. Carriemore car trans, red cab, blue trailer, Corgi decals on side, two or four cars as required, Corgi hydraulic jacks fitted on this wonderful model allowing the top deck to lower slowly like a real transporter, 14/11d, £1.8.11d, 263 mm	£30	£20	£15	£12
No 1119 H D L Hovercraft SR/N1. Blue, white, grey, central air intake, propulsion ducts, control cabin, seats, elevans, rudders which are all carefully and accurately reproduced, red and yellow decals, 8/6d, 120 mm, An exact replica in miniature of the first of the new land sea air vehicles. Supported on cushions of air, opening up a new era in long distance travel.	£15	£12	£9	£7
No 1107 Euclid T C 12 Tractor with Dozer Blade. Red, driver, grey trim, black rubber tracks, grey blade, lever operations, 13/6d, 159 mm	£18	£15	£12	£10
No 55 Fordson Power Major Tractor. Blue, grey, orange, 6/9d, 91 mm	£15	£12	£10	£8
No 50 Massey Ferguson 65 Tractor. Red, salmon, driver, 4/7d, 79 mm	£10	£8	£5	£3
No 54 Fordson Power Major with Roadless Half Tracks. Blue, dark orange, grey trimming, 6/9d, 91 mm	£15	£12	£10	£7
No 18 Gift Set. Fordson power major tractor, blue and orange, red and yellow; grey trim, four furrow plough, 9/4d	£18	£15	£12	£10
No 7 Gift Set. Massey Ferguson tractor, red, salmon and yellow, tipper trailer, 7/9d	£18	£15	£12	£10
No 22 Agricultural Gift Set. Massey Ferguson combined harvester, driver, red, silver and orange, Massey Ferguson Tipper Trailer, red, orange, skip, milk churns, Massey Ferguson 65 Tractor muck shifter, driver, Land Rover, seats, steering wheel, glido spring suspension platform trailer, milk churns load detachable, Fordson Power Major Tractor, steering wheel control, vibrating exhaust, plough lifting mechanism, driver, 4 furrow plough, 3 drivers for combine harvester and two tractors, All with proper decals and correct company colours, 50/-d, This gift set comes in a special polyfoam packing within a presentation box. The polyfoam protects every item individually and enables you to keep your models safely. This magnificent gift set is one of the best ever produced.	£60	£50	£20	£15
No 605 Silverstone Club House & Timekeepers Box. Green, cream, 5/6d, 191 mm	£1.50	£1	75p	50p
No 604 Silverstone Press Box. Green, yellow, various decals for building, 3/-d, 146 mm	£1.50	£1	75p	50p
No 603 Silverstone Pits. Decals in red, green, brick and white, 5/-d, no Corgi layout on the racing scene is complete without one of the easy to construct Corgi kits. They are easily joined together, 216 mm	£2	£1	75p	50p
No 25 Gift Set. Shell or BP garage layout. This wonderful realistic gift set includes a Shell or BP service station with all the forecourt accessories plus garage attendants complete with the following: Two 606 Lamp Standard Sets, 602 AA & RAC Telephone Boxes, Three 601 Batley Leofric Garages, 224 Bentley Continental, Austin 7 Saloon, Chevrolet Corvair, Ford Consul Classic; Ford Zephyr motorway patrol car, to complete the set a coloured background is provided on which all the components can be positioned, all transfers with adhesive paints and brush, 30/-d, 457 x 762 mm	£50	£40	£10	£5

CORGI

	Mint boxed	Mint unbxd	Good cond	Fair cond
No 602 RAC & AA Telephone Boxes. Authentic colours, decals, 3/-d, 64 mm	£1	75p	50p	25p
No 1103 Euclid T C 12 Twin Crawler Tractor. Light green, grey trimming, 10/-d, 111 mm	£15	£10	£7	£5
No 2S Gift Set. Land Rover with Rice's Pony Trailer, rich brown, cream covers, blue wheels, grey tyres, horses; seats, steering wheel; glidomatic suspension, 18/6d, 266 mm	£25	£15	£12	£10
No 1130 Chipperfields Circus Horse Transporter. Red, blue, correct Chipperfield decals, 18/6d, 260 mm, This super model includes all the fascinating features of the real Chipperfields circus and all the following items on this page are also in the proper colours designed by craftsmen of a very high standard. The authenticity of their work has been praised all over the world.	£40	£25	£10	£7
No 426 Circus Booking Office. 5/9d, 91 mm	£6	£5	£3	£2
No 607 A Corgi Kit. Circus elephant, transport cage, 5/9d, 76 mm	£6	£5	£3	£2
No 1121 Circus Crane Truck. 11/8d, 200 mm	£20	£15	£5	£3
No 1123 Circus Cage with Animals. 11/8d, 127 mm	£10	£8	£6	£4

No 23 Circus Gift Set. The items included in this gift set are the Circus Crane truck, 2 Circus animal cages, circus land rover, seats, steering wheel, glido suspension, platform trailer, circus elephant in transport cage, booking office with glido suspension, 2 lions, 2 polar bears, it is in a special polyfoam packing in a beautiful presentation box, 50/-d

	Mint boxed	Mint unbxd	Good cond	Fair cond
	£75	£50	£20	£15
No 12 Gift Set. Circus crane truck, circus cage, animals, Packed in special presentation box, 23/4d	£30	£25	£12	£10
No 19 Gift Set. Land rover, elephant in cage on trailer, packed in presentation box, 11/6d	£30	£25	£12	£10

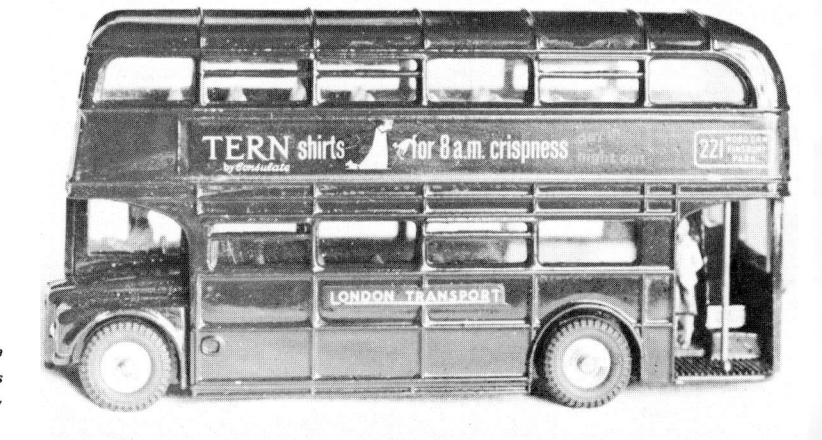

A Corgi version of a London Transport double decker bus in the familiar red livery

Corgi classics

Nearly every toy firm in the world has at one stage or another, introduced an old vintage car series of one type or another. The Corgi Classic ranks high among vintage cars.

	Mint boxed	Mint unbxd	Good cond	Fair cond
No 901 1915 Model T Ford. Black, driver, passengers, gold trim, hood folded down, Tin Lizzie	£40	£25	£12	£10
Model T Ford. Blue, lady driver, gold trim, orange wheels, black hood	£40	£25	£15	£12
No 900 1927 Bentley. Green, white driver, silver wheels, Union Jack decal, 3 litre Le Mans winner	£45	£25	£12	£10
No 9021 1910 Daimler. Red, four passengers, silver trim, colours may vary	£45	£25	£15	£12
No 9032 1910 Renault. Mauve, black, gold trim, open windscreen, also in yellow, black, rare model	£35	£25	£15	£12
No 9001 Bentley. Red, white driver, black hood	£30	£25	£20	£15
Bertie Wooster's Car & Jeeves. Green, black	£20	£15	£12	£10
Rolls Royce Silver Ghost. Silver, gold, very rare	£30	£25	£20	£15

The model T 1915 Ford in the Corgi Classics series

The 1910 Daimler complete with passengers in the dress of the period

Spot on

Spot On models were made in a modern plant in Belfast, Northern Ireland. A team of highly skilled designers and production engineers created Spot on models. Descriptions also include some background information of the cars on which models are designed.

	Mint boxed	Mint unbxd	Good cond	Fair cond
No 100 Ford Zodiac. Two tone blue, fawn wings, 2½-litre engine, 6 cylinders, 111 mm, This car was the pride of the Ford range, it was renowned for its vivid and quick acceleration and high cruising speed, Many of the optional extras quoted by other makers are standard fittings on the Zodiac	£10	£7	£5	£3
No 101 Armstrong Siddeley Sapphire 236. Golden yellow, black roof, silver trim, 6 cylinder engine, 2-3-litre, fitted with twin carburetta set, 108 mm, A specialist built car with a very smooth performance, new, 'no clutch' Sapphire claims to be a car that thinks for you, top speed of 85 miles per hour and over 100 mph	£20	£10	£7	£5
No 102 Bentley 4 door Sports Saloon. Silver grey lower, red top, bonnet, 127 mm, The most luxurious sports saloon of it's day, developed from a long line of racing cars, this is the best version ever, the engine and all other specifications are identical to the Rolls Royce, a real winner	£20	£10	£7	£5
No 103 Rolls Royce Silver Wraith. Deep maroon lower, silver grey top, bonnet, silver trim; seats five; 6 cylinder, 4,887 cc engine, overhead inlet, side exhaust valves, 140 mm, still regarded as the world's best car, the standard by which all others are judged, touring limousine	£30	£15	£10	£8
No 104 MGA Sports Car. Red, blue silver trim, yellow interior, 96 mm, This favourite of the enthusiast who drives hard and well, its four cylinder 1,489 cc engine develops 72 bhp at 5,500 rpm and in 1956, the MGA took class F records in the USA, a specially modified car, maintaining and doing an average speed of 141 mph for 12 hours	£15	£10	£7	£5
No 105 Austin Healey '100-Six'. Green, silver trim; 86 mm, This car is classed as an occasional 4-seater sports tourer, this vehicle first became available to the motorist in September 1956, Powered by a 6 cylinder 2,639 cc engine, giving 102 bhp at 4,600 rpm, it has proved beyond all doubt that it is very capable of prolonged high speed motoring in excess of 100 mph; this car has won top awards overseas	£15	£10	£7	£5
No 107 Jaguar XK SS. Red, silver trim, yellow interior, 91 mm, the fastest sports car ever made, a modified version of the 'D' type so successful at Le Mans, capable of over 170 mph, fitted with disc brakes on all wheels, the XKSS represents another triumph for British engineering skill	£15	£10	£7	£5
No 108 Triumph TR3. Green, silver trim, orange interior, 89 mm, still one of our best production sports cars, over 100 mph, disc brakes, good handling qualities coupled with economy, deservedly makes the TR3 popular with the enthusiasts	£15	£10	£7	£5
No 112 Jensen 541. Yellow, black roof, silver trim, 108 mm; this individually built, glass fibre bodied sports saloon, is one of the finest automobiles available today, maximum speed 125 mph, servo assisted disc brakes, powerful 4,000 cc engine, enables it to more than hold its own with its continental competitors	£20	£12	£10	£7

SPOT ON

	Mint boxed	Mint unbxd	Good cond	Fair cond
No 113 Aston Martin DB3. Dark green, silver trim, 105 mm, developed from a long line of racing cars, and undisputed leader of its class, the DB3 has a tubular chassis, an alloy body and disc brakes, and is capable of over 120 mph, probably the finest sports saloon available today, and one that Britain can justifiably be proud of	£15	£12	£10	£7
No 114 Jaguar 3.4 litre. Maroon, 108 mm, beautifully bodied production sports saloon, developed from a long line of famous racing cars, graceful, capable of over 120 mph safely, with its disc brakes and lion hearted engine, this car is deservedly one of the most popular of its type	£15	£12	£10	£7
No 115 Bristol 406. Pale green, silver trim, 115 mm, a welcome addition to the range of high quality Sports Saloons for the enthusiast, manufactured by Automobile division of the Bristol Aircraft Co, the 406 is engineered to the same high standards, capable of over 100 mph, fitted with disc brakes on all wheels, the 406 represents British craftsmanship at its best	£15	£12	£10	£7
No 118 BMW Isetta. Lemon, silver trim, 57 mm, this deservedly popular bubble car, many thousands of which are on the roads today, has been hailed as the answer to the traffic and parking problems in town, its nippy performance and economical running costs have made it a firm favourite throughout Europe	£20	£12	£5	£3
No 119 Meadows Friskysport. Red, grey soft top; this popular plastic bodied minicar, styled by Italy's Michelotti, combines high performance with maximum economy, well sprung easy to park and capable of 60 miles per gallon, the Frisky is the complete answer to low priced and comfortable motoring	£12	£10	£7	£5
No 120 Fiat Multipla. Medium green, 91 mm, this novel Italian six seater is one of the most interesting in the current range of small cars, roomy, comfortable, solution to family transportation, easy to park, economical to run	£12	£10	£7	£5
No 131 Goggomobil Super. Red, mustard roof, 70 mm, this German designed and produced Coupe has full sized car lines and comfort and combines high cruising speeds with minimum fuel consumption, it has many				

The Jaguar S model

The Vauxhall Cresta

	Mint boxed	Mint unbxd	Good cond	Fair cond
novel technical features, including an electro-magnetic gearbox and a unique rear engine suspension unit, add this to the spacious and well designed body and it is easy to see why the Goggo has been so successful	£20	£12	£10	£7
No 210 Morris Mini Minor Van. Mustard, 79 mm, another winner from the BMC stable, a zippy, roomy and economical 45 mpg baby, completely new in every way, front wheel drive, transverse engine, host of ingenious design features	£20	£12	£10	£7
No 155 Austin Taxi. Black, silver trim, yellow seats, 5 seater, 117 mm, handsome vehicle will be seen in increasing numbers as it replaces the older types, powered by 2.2 litre diesel engine with automatic transmission	£25	£12	£10	£7
No 157 Rover 3 Litre. Pale green with silver body, 108 mm, an entirely new body and engine from one of England's leading manufacturers, craftsman built, luxuriously appointed, superbly finished, Rover 3 litre represents British engineering skill at its best	£20	£15	£7	£5
No 165 Vauxhall Cresta. Maroon, silver trim, 115 mm; the pride of the Vauxhall range, transatlantic in its styling, as befits a subsidiary of General Motors, the Cresta represents a satisfying and harmonious blend of the best British and American automobile practice	£20	£15	£9	£7
No 166 Renault Floride. Pillar box red, silver trim, 102 mm, developed from famous Dauphine but fitted with a more powerful engine, the new Floride is sure to become as firm a favourite as its predecessor. Available as an open tourer or saloon, the elegant Floride has few rivals in its class	£20	£15	£9	£7

SPOT ON 119

	Mint boxed	Mint unbxd	Good cond	Fair cond

No 191 Sunbeam Alpine. Cream, black roof, 96 mm, An entirely new successor to the world famous Sunbeam line; low, sleek, terrific performance, has become a firm favourite with enthusiasts here and in USA — £15 £10 £7 £5

No 154 Austin A40. Bright red, black roof, silver trim; 89 mm, another winner from the Austin stables, the Farina styled body, with the well tested A35 engine and transmission unit, makes the zippy A40 a firm family favourite, low running costs, economical fuel consumption, plus its reasonable price, gives the A40 the popularity it so well deserves — £15 £10 £7 £5

No 211 Austin Baby Seven. Grey/blue, 70 mm, companion to the mini minor, in all respects technically identical, although differing in the radiator grille, the new Austin Seven is destined to become as famous as its earlier but equally admired namesake — £15 £10 £7 £5

No 213 Ford Anglia. Medium blue bottom, pale blue roof; 96 mm, an entirely new model and easily recognized by the cut back rear window, new Anglia is a brilliant answer to the motoring problems of the average motorist, and has already proved a huge success — £20 £15 £10 £7

No 215 Daimler SP250. Red, yellow interior, 97 mm, new revolutionary sports car from one of our oldest manufacturers fibre glass bodied, a V8 engine with disc brakes on all wheels, the SP250 with top speed of 120 mph, will be a worthy challenger to the best of the Continental sports cars — £20 £15 £10 £7

No 207 Wadham Ambulance. Cream, blue flashes, 115 mm, this handsome vehicle has an all fibre glass body and fitted to a BMC petrol or diesel engine and chassis unit, design features include a wrap around windscreen and an ingenious interior arrangement - a fine example of British specialised vehicle design — £30 £20 £12 £10

No 100 S/L Ford Zodiac. Red, cream, silver trim, the economical 2½-litre, 6 cylinder engine will transport six passengers with the comfort and pace never previously associated with vehicles in its price range, the Zodiac, one of the famous Three Graces, is backed by the Ford Spares and Service organisation unquestionably the finest in the world — £20 £15 £9 £7

106A/OC Austin Type 503 Normal Control. Blue, green load, 236 mm, this popular vehicle of which many thousands are in daily use, can be obtained as an articulated prime mover or rigid chassis vehicle — £25 £20 £12 £10

No 110/3 AEC Mammoth Major 8. Red, British Road Service decals, 210 mm, The Mammoth Major built by the AEC Co makers of many of London's bus engines and chassis, is the latest development of many years operation and experience, successfully used by many operators hauling loads from bricks to bulk liquids, the Major 8 has won many admirers through its reliability, toughness and economy — £40 £25 £12 £10

No 161 Long Wheel Base Land Rover. Grey, lemon roof; 108 mm, a variation of the well known land rover of which over 250,000 have been produced, its rugged construction, powerful engine and four wheel drive, enable it to go anywhere, the illustration shows a tropical version with a double roof for heat protection — £30 £20 £12 £10

No 110/4 AEC Mammoth Major 8. Red cab, green tanker, 210 mm, the Mammoth Major built by the AEC Co makers of many of London's bus engines and chassis is the latest development of many years operation and experience. Successfully used by many operators hauling loads from bricks to bulk liquids, the Major 8 has won many admirers through its reliability, toughness and economy — £30 £20 £12 £10

No 111A/OT Ford Thames Trader. Red lorry, green load, 219 mm, one of the smartest and most modern of the commercial vehicles on the road today, it is available in many body/load combinations — £30 £20 £12 £10

No 106A/1 Austin Type 503 Normal Control. Bright yellow, 236 mm, this popular vehicle, of which many thousands are in daily use, can be obtained as an articulated prime mover or rigid chassis vehicle, if articulated it can be fitted with the standard Scammell coupling gear to haul a wide variety of trailers — £25 £18 £12 £10

No 109/2 ERF 68G. Green, black roof, base, 210 mm, this modern 8 wheeler, the first to use the wrap around windscreen, is deservedly popular with hundreds of operators, powerful, robust and reliable, the 68G repre-

SPOT ON

	Mint boxed	Mint unbxd	Good cond	Fair cond
sents one of Britain's major contributions to heavy road transport	£40	£15	£12	£10
No 110/2B Mammoth Major 8. Red; 210 mm, built by AEC Co, makers of London's bus engines and chassis, used for hauling loads from bricks to bulk liquids	£40	£15	£12	£10
No 158A/02 Bedford 10 Tonner. Red body, green cab, BP Shell petroleum decals on sides of doors and tanker, front of cab, 199 mm, this powerful diesel engined vehicle can be obtained as an articulated or rigid chassis version, this is an articulated prime mover fitted with Scammell coupling gear, hauling a Shell 2,000 gallon tanker	£40	£15	£12	£10
No 139 Eccles E 16 Caravan. Blue, white roof, 146 mm, a typical example of modern 4 berth caravan design, the E 16 may be seen on many coastal caravan sites, comfortably furnished, it has provided happy holidays for its owners and their families	£15	£10	£7	£5
No CB106 Four Wheel Trailer. Yellow, grey chassis, 174 mm, this is the spot on version of the versatile four wheel trailer, specially designed to be coupled on to either your ERF 68G or Mammoth Major 8, this combination will give you an exact replica of the big road haulage unit	£10	£6	£4	£2
No B23 Riva Speedboat (Ariston). Brown, fawn, green, blue, yellow trim, 156 mm, classic, elegant, swift, here is Ariston contructed in accordance with the best Italian tradition, easily manoeuvred with exceptional sea qualities, more solid than any other motor boat of same style, with driver	£15	£10	£7	£5
No 109/3 ERF 68G Flat Float with Sides. Orange, black, grey chassis, 210 mm, this modern 8 wheeler, the first to use the wrap around windscreen, is deservedly popular with hundreds of operators, powerful, robust and reliable, the 68G represents one of Britain's major contributions to heavy road transport	£30	£20	£12	£10
No 117 Jones Crane. Red, yellow, grey, black, tow hook, 289 mm, a great member of the Jones mobile cranes, the KL 10-10 is equipped with a single 125 hp diesel unit and can travel at road speeds up to				

	Mint boxed	Mint unbxd	Good cond	Fair cond
30 mph, completely mobile and with a capacity of 12½ tons, this model is a magnifent example of British engineering craftsmanship, and another great triumph for its makers, K&L Steelfounders and engineers Ltd, who are members of the George Cohen 600 group	£30	£20	£12	£10
No 110/2 AEC Mammoth Major. Dark blue, silver grey, silver trim, 210 mm, the Mammoth Major, built by the AEC Co, makers of many of the London buses for their engines and chassis, is a typical example of their skill after many years of development, and experience; it is tough and has many admirers	£30	£25	£13	£10
No 111/A1 Ford Thames Trader. Pink, yellow, silver trim, silver or grey wheels, 219 mm, one of the smartest wagons ever made, a commercial to be proud of, available in many body/load combinations, was available as an articulated prime mover to haul a wide range of things	£30	£20	£12	£10
No 123 Bamford Excavator. Yellow, red, red lines, JCB decals, 272 mm, the JCB Hydra Digga is made of tubular and boxed steel section, tremendously robust construction and made to withstand all shocks and stresses, it is powered by a 51.8 hp diesel engine and operated by five rams, these are very specially designed and produced by JCB, fitted with 750x16 front, and 14x30 rear wheels and tyres, electric lighting and starting; handbrake and all weather cab	£25	£20	£12	£10
No 158 A/3C Bedford Low Loader with Cable Drum. Red, silver trim, dark blue wheels, 270 mm, this great general purpose haulage vehicle which can be obtained as an articulated or rigid chassis version has Scammell coupling gear, hauling a low loader with cable drum winch unit	£40	£25	£15	£10
No L135 GP14ft Sailing Dinghy. Blue, white, 117 mm, undoubtedly one of the most successful designs ever made, the GP 14' is suitable for racing rowing, or for the use with an outboard motor. Designed by Jack Holt, a great favourite with the yachting world, many hundreds have been built and may be seen at any sailing club	£10	£8	£6	£4
No 145 London Routemaster Bus. Red, correct bus decals, driver, signs and adverts may vary, 198 mm,				

SPOT ON

	Mint boxed	Mint unbxd	Good cond	Fair cond
this magnificent vehicle is one of the most well loved in the world, more powerful and comfortable than most and was a worthy successor to the long line of London Buses, whose lineage goes back to the pre 1914 Horse drawn buses, very rare, collectors gem	£40	£30	£15	£10
No 156 Mulliner Luxury Coach. Silver grey, dark blue, silver trim, 207 mm, a typical example of a modern luxury coach, this vehicle boasts advanced body styling plus the last word in comfort for its passengers, and is a far distant cry from its ancestor, the old English Stage Coach, this bus like many other models of this type are scarce and rare, they will be almost unobtainable for the future, one of the best investments a collector can have, perhaps you may think that the price quoted is high, but this is because of the great demand; they can hardly be seen, and soon the only place one will see one will be in a museum, all buses are sought after so if you have any of these, or the heavy goods vehicles, please get the correct market price for them, you can always consult me for advice on any matter to do with rare items	£50	£40	£15	£10
No 122 United Dairies Milk Float. Red, white or blue, white decals, black trim, 97 mm, manufactured by the United Dairies Ltd, this modern electric powered silent delivery vehicle may be seen in almost every London district and at long last gave the horse its well earned rest	£25	£20	£7	£5
No 137 Massey Harris Tractor. Red, blue, silver trim, 79 mm; this powerful modern tractor is in daily use by thousands of farmers all over the world, it boasts of an efficient haul, Hydraulic system for raising and lowering the various implements, this spot on model is designed to pull a range of things, these are specially made by Spot On with great care	£25	£20	£10	£7
No 116 Caterpiller Tractor D9. Mustard or dark brown, white or blue blade, driver, white tracks, 153 mm, one of the most impressive tracked vehicles ever produced, this Cat D9 can be seen in action on many of the worlds major construction projects; powered by a turbocharged 286 hp diesel engine and equipped with a wide variety of useful attach-				

	Mint boxed	Mint unbxd	Good cond	Fair cond
ments, the Cat D9 is capable of sustained round the clock working, under the toughest conditions, the blade is raised and lowered by a powerful engine driven winch	£25	£13	£10	£8
No 0 Presentation Set. Austin prime mover; Flat float with sides, Ford Zodiac, Rolls Royce Silver Wraith, Aston Martin DB3, MGA Sports car, 255 x 204	£90		£50	
No 1 Presentation Set. Ford Zodiac; Armstrong Siddeley Sapphire, Rolls Royce Silver Wraith, MGA Sports Car, road signs, zebra crossings, white lines, 408 x 370 mm	£90		£40	
No 2 Presentation Set. ERF 68G flat float with sides, Bentley four door saloon, Armstrong Siddeley saloon; Austin Healey sports car, road signs, crossings etc, 408 x 395 mm	£90		£45	
No 3 Presentation Set. Thames trader with art, flat float Sapphire 236, MGA sports car, Triumph TR3 Jensen 541, Aston Martin DB3, Jaguar 3.4, Beacon straights road signs, 408 x 433mm	£100		£65	
No 4 Presentation Set. Austin Prime mover with flat float, ERF 68G with flat float, Rolls Royce Silver Wraith, Ford Zodiac, Jensen 542, Jaguar XK SS, Straight white lines, road signals, belisha beacon, 408 x 433 mm	£100		£75	
No 0A Presentation Set. Austin Taxi cab, Goggomobile super, Bristol 406, Jaguar XKSS, Rover 3-litre, Fiat multipla, ERF 68G with flat float, 286 x 325 mm	£100		£70	
No 2A Presentation Set. Bedford 10 Tonner, artic, 2,000 gallon tanker, LBW Land Rover, Austin Taxi, Austin Healey Meadows frisky sport, Vauxhall cresta, Rover 3-litre, ERF 68G flat truck, flat float with sides, 312 x 395 mm	£150		£85	
No 4A Presentation Set. BP filling station, straights, dotted line, zebra crossings, road signs, accs for road up or under construction, LBW Land Rover, Sunbeam Alpine, hardtop, Vauxhall cresta, AEC Major 8, brick load, 542 x 408 mm	£75		£65	

Lone star

In addition to producing some wonderful railway models and accessories Lone Star turned out some beautiful die-cast models. Although their range was limited there are still a few available. I am listing a set that is complete and called the 19 Fantastic Flyers. There are of course other models apart from this set.

	Mint boxed	Mint unbxd	Good cond	Fair cond
No 9 The Maserati Mistral. Gold, silver trim	£5	£4	£3	£2
No 10 Jaguar Mk 10. Correct company colours	£5	£4	£3	£2
No 11 Gran Turismo Coupe. White, silver trim	£5	£4	£3	£2
No 12 Chrysler Imperial. Black, silver trim	£5	£4	£3	£2

Lone Stars version of the Cadillac

	Mint boxed	Mint unbxd	Good cond	Fair cond
No 14 Ford GB Zodiac Mk III Estate. Opening doors	£5	£4	£3	£2
No 15 Volkswagen Microbus. Yellow, black decals	£6	£5	£3	£2
No 16 Motorway Police Car. White, opening doors	£5	£4	£3	£2
No 17 Mercedes Benz 220SE. Black, silver trim	£6	£5	£3	£2
No 18 Ford GB Corsair. Blue, white trim	£5	£4	£3	£2
No 19 Volvo 1800S. Red, black, opening doors	£5	£4	£3	£2
No 20 Volkswagen Ambulance. Red X decals, opening doors	£6	£5	£3	£2
No 21 Fiat 2300S. Coupe. Red, opening doors	£5	£4	£3	£2
No 22 Rolls Royce Silver Cloud III. Silver grey, black trim	£6	£5	£3	£2
No 23 Alfa Romeo Giulia 1600 Spider. Silver, black	£5	£4	£3	£2
No 27 Ford Taunus 12M. Black, Made Germany	£5	£4	£3	£2
No 28 Peugeot 404. Grey, blue decals	£5	£4	£3	£2
No 32 Fire Chief's Car. Red, light on roof	£6	£5	£3	£2
No 36 Lotus Europa GT. Metallic grey, black trim	£5	£4	£3	£2
No 37 Ford GT 40. White, red trim	£5	£4	£3	£2
London Routemaster Bus. Red, driver, conductor	£10	£5	£4	£3
Starters Set. 4284 mm flyway flexible track, loop the loop fittings; anchor clip, 1 flyers car, flyers club badge	£10	£6	£5	£3
Double Racing Set. Gay colours; 7344 mm flyway flexible plastic track, twin loop the loop fittings, 2 anchor clips, 2 flyers cars, 1 flyers club badge	£14	£10	£6	£4

Britains

No collectors guide would be complete without reference to the firm of Britains. They have specialised in toy lead soldiers and beautiful coaches and other similar items for many years.

	Mint boxed	Mint unbxd	Good cond	Fair cond
No 9401 Her Majesty's State Coach. Blue, gold, dark brown, red trim, 8 horses, white, gold, red trimming, 4 riders, red, gold jackets, black caps, black boots, passengers, model finished in die cast metal and bronze, finest in the world, 50/-d	£200	£125	£75	£50
No 9402 State Open Road Landau. Decorative open Landau, 6 white horses, black, gold trimming, Queen, consort, drivers, 2 footmen; 50/-d	£500	£400	£250	£100

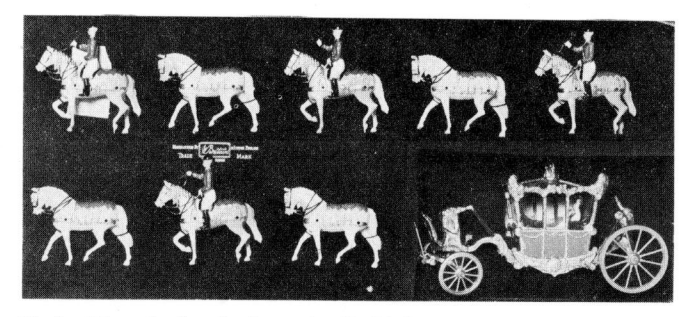

The Royal Coronation State Coach as produced by Britains

124 BRITAINS

Minic was created by Triang Products and their models are of the highest quality.

	Mint boxed	Good cond
No 1558 Mercedes Benz 300SL. White, red top, No 52, 19/6d	£5	£2
No M1573 Aston Martin DB4 GT. Gold metallic, No 8, 16/1d	£5	£2
No M1559 E Type Jaguar. Green, No 7, 19/6d	£7	£4
No M1574 Porsche Carrera GT. Red; No 4, 16/11d	£5	£3
No M1568 Jaguar 34 Saloon. Silver continental, 19/6d	£10	£5
No M1543 Humber Super Snipe. Green, silver trim, 19/6d	£7	£4
No M1555 Trailer with Boat. Blue, white, 3/6d	£3	£1
No M1542 Jaguar 3.4. Metallic blue, silver trim, 19/6d	£7	£4
No M1567 Jaguar 3.4 Saloon. Green, British racing, 10, 19/6d	£10	£5
No M1552 Police Car. Black, warning light, 21/6d	£10	£5
No M1541 Rolls Royce Silver Cloud. Metallic blue, 19/6d	£15	£7
No M1569 Conqueror Tank. Brown, letter 7, 19/6d	£20	£10
No 1556 Mercedes Benz 220S. Black, silver, grille, 19/6d	£10	£5
No M1553 Caravan. Green or white, 2/11d	£2	£1
No M1551 Shell Oil Tanker. Blue, fawn, white, 25/-d	£15	£7

	Mint boxed	Good cond
No M1547 Bedford Lorry with Container Load. Grey, blue, 25/-d	£20	£10
No M1546 Bedford Lorry and Bale Load. Mustard, green, 25/-d	£15	£10
No M1545 Double Decker Bus. Red, John Player decals on the side; destination on front, rear, London transport, 25/-d	£40	£25
No M1563 Steam Lorry with Smoke. Blue, coal decal on front, 32/6d	£30	£20
No M1564 Steam Lorry without Smoke. Blue, decal on front, 21/-d	£20	£10
No M1565 Breakdown Lorry with Towing Light. Blue, yellow, day and night service, Jackson Motors decals on sides, 25/-d	£15	£10
No M1550 Fire Engine. Red, Kent fire brigade, flashing light, ladder, 25/-d	£15	£7
No M1548 Car Transporter. Yellow, blue, grey loader, cars can be driven on to both decks for distribution by transporter, 35/-d	£20	£15
No M1544 Streamlined Coach. Fawn, purple flash; silver trim, 25/-d	£25	£15
No M1554 Trailer. Yellow, for use with commercial vehicles, 2/11d	£3	£1
No M1581 Aston Martin DB6. Green, silver trim, No 4, 21/6d	£6	£3
No M1582 Jaguar E Type Fastback 2+2. Red. silver trim, 10, 21/6d	£6	£3

Minic

	Mint boxed	Good cond
No M1556 Mercedes Benz 220S. Light green, No 22, 21/6d	£6	£3
No M1576 Ferrari 500 Superfast. Red, silver trim, No 21, 21/6d	£6	£3
No M1577 Chevrolet Corvette Stingray. Gold, silver trim, No 35, 21/6d	£15	£7
No M1549 Firechief Humber. Red, Fire Chief decal, 19/6d	£10	£4
No M1552 Police Jaguar. White, police decal, warning light, 19/6d	£10	£5
No M1545/G Double Decker Bus. Green, Green Line decals, 25/-d	£25	£15
No M1801 Filling Station, 229 x 76 mm, 9/11d	£5	£2
No M1814 Motel Chalet & Garage. Green, yellow, 229 x 168 mm, 12/3d	£7	£4
No M1804 Fire Station. drive-in/drive-out, 229 x 156 mm, 14/-d	£6	£3
No M1810 Dunlop Footbridge. Grey, Famous Le Mans bridge, 7/-d	£3	£1
No M1706 Set of 6 Flags. All International Standards; 1/11d	£1	50p
No M1707 Length of White Fencing. 153 mm, 2½d	30p	10p
No M1809 Racing Pits. Important trackside items, 8/11d	£2	£1

	Mint boxed	Good cond
No M1525 Gran Turismo Racing Car Set. Aston Martin DB4 GT, Porsche Carrera GT, over and under figure of 8 track and 2 speed controllers, 816 x 663, Targa Florio here, The Aston and Porsche are miniature counterparts of these two fabulous marques; 79/11d	£25	£15

No M1526 The Trident. Mercedes Benz, 300 SL, E Type Jaguar, 3.4 Jaguar, extra large over and under

The Royal Mail van (Minic)

126 MINIC

	Mint boxed	Good cond
figure of 8 track with 3 speed and direction controllers, 1224 x 969 mm, 177/6d	£30	£25
No M1522 The Europa Set. Mercedes Benz 300SL, E Type Jag, large figure of 8 track with hump back bridge, two speed direction, controllers, 6350 x 663 mm, E type and Merc, magic names, superb lines, 113/6d	£25	£15
No M1504 The London Set. Rolls Royce, Humber Snipe, 8 piece oval track, two speed, direction controllers, 816 x 508 mm, opulent Rolls, sedate Humber both gleaming with plate accessories suggesting first nights and tycoons, 85/11d	£25	£15
No M1511 The Trunker Set. Bedford Lorry, container load; Shell oil tanker, 8 piece over track, two speed, directional controllers, 816 x 508 mm, no need for the Minister of Transport to inspect these two Minic Motorway vehicles because they are perfect in every way, 94/6d	£35	£25
No M1512 The Traveller Set. Double decker bus; luxury coach, large figure of 8 track, cross roads, two speed and direction controllers, 1122 x 864 mm; 113/6d	£45	£25
No M1812 Automatic Starting Gate. As Union Jack drops the cars surge forwards, and in the event of a pile up, Marshall drops red danger flag which automatically checks all cars, 12/6d	£4	£2

Minics smallest clockwork car

The introduction of the Italian 'Rio' models was an instant success. They were expensive but unique. Each of the Rio veteran cars series had a different style and told its own story. Rio models are already becoming scarce, like the No 20 'The Fiat Autobus', and as only limited numbers of models are made each time a new addition is introduced they are very quickly picked up.

	Mint boxed	Mint unbxd	Good cond	Fair cond
No 1 Italia Targa Florio 1906. Red, blue, winner of the madonia circuit race in 8 hrs 32 mins, over the roughest course average speed, 27.5 mph, 4 cyl, 7433 cc 45 hp	£10	£7	£5	£3
No 2 Italia Peking Paris Raid 1907. Light blue, orange wheels; piloted by S Borghese, car covered 9600 miles in 61 days from Peking to Paris, 4 cyl 40 hp engine	£6	£5	£3	£2
No 3 Fiat 501 Tipo Sport 1919-26. Red; silver trim; black wheels, black interior, motor 4 cyl 26 hp, 3000 tpm, top speed 50 mph, open sports	£10	£8	£5	£3
No 4 Fiat 501 Torpedo Lusso 1919-26. Green, silver trim, black hood, wheels, 4 cyl, 22 hp, 1460 cc (65 x 1000 engine)	£10	£7	£5	£3
No 5 Alfa Romeo Grand Prix P3 1932. Red, grey trim, silver trim, average speed 120 mph, 8 cyl V engine 180 hp, 5400 tpm, 1st prize race of Italy with Nuvolari driving, first one seat racer	£7	£6	£4	£3
No 6 Fiat Model 0 1912. Red, black, fawn top, gold trim; motor 4 cyl, one block 12/15 hp, 35 mph, most famous early Fiat in the world, top prizewinner	£10	£8	£6	£4
No 7 Fiat Model 0 1912. Yellow, black, gold trim, Spider Coupe with rumble seat, developed after the great Ford, is same as 6 engine regards power	£8	£7	£5	£3

Rio

	Mint boxed	Mint unbxd	Good cond	Fair cond
No 8 Issotta Fraschini Typ 8a 1924. Dark purple, black roof, silver trim, this is the Italian Rolls of 1924, on this doors open to show jumpseats, spring, bmps	£10	£9	£7	£5
No 9 Issotta Fraschini 1924 Open. Rich ruby, black roof, silver trim; 100 mph; the famous town car, body by Sala, 100/126 hp, wire wheels by Rudge, a real beauty	£12	£10	£7	£5
No 10 Bianchi Landaulet 1909. Pink, blue; gold trim; eng 4 cyl, 25 mph to 50 mph, 25 hp, 1250 rpm	£9	£8	£6	£4
No 11 Bianchi Landaulet. Deep green, black; gold trim, open back, engine details as No 10	£10	£9	£7	£5

An alternative three wheeler car from Rio

	Mint boxed	Mint unbxd	Good cond	Fair cond
No 12 Fiat Model 0 1912. Green, black/blue, gold trim, orange wheels, open sports, the spider	£8	£7	£5	£3
No 13 Fiat 508 1932/37. Blue, black, silver trim, pink wheels, the famous balilla 955 cmc, 4 cyl in line	£10	£9	£6	£5
No 14 Fiat Type 2 1910-20. Rich dark blue, green, orange trim, gold, in line 4 cyl, 28 hp engine, 4 speeds forward with a very unusual windshield, a beauty	£10	£9	£6	£4
No 15 Issotta Fraschini Type 8a. Yellow, red line trim; blue trim, black wheels, white tyres, spider 1924 with body by Sala of Milan, engine 8 cyl, 120 hp, 100 mph	£12	£10	£7	£5
No 16 Chalmers Detroit 1909. Green, blue, gold trim, orange wheels, 4 cyl, 3/6 liters in line	£7	£6	£4	£2
No 17 Mercedes Benz 1909. White, red interior, silver trim; orange wheels, 4 cyl in line, 4084 cmc model has chain transmission, one of the best ever made	£10	£9	£7	£5
No 18 Bianchi 15-20 v 106. Yellow back, green chassis; black, orange, gold trim, coupe de ville 8 cyl in line engine cmc 4939 the buggie on chassis	£12	£10	£7	£5
No 19 Alfa Romeo 6 C 1750-1932. Cream, rich wine, valve in head 55 hp, top speed, on this model the cover of the engine opens to show it in complete detail	£15	£12	£10	£7
No 20 Fiat Autobus 1915. Golden yellow, decals along top of bus, Firenze Poggibonsi, Volterra, single block 4 cyl, 30 hp engine, a great rare collectors gem	£25	£20	£15	£10
No 21 Large Mercedes 1938 Open. Greenish grey body, black hood down at back; red interior, silver trim;				

RIO

	Mint boxed	Mint unbxd	Good cond	Fair cond
silver, black wheels; lift off bonnet	£15	£12	£10	£8
No 22 Large Mercedes 1938. Rich ruby, metallic silver blue trim, black wheels; black roof; wine interior, engine 8 cyl in line, head valves cmc 7700, 110 mph	£15	£14	£12	£10
No 23 Fiat 60 cv 1905. Blue, gold trim, grey roof, red interior; 6 cyl engine in line, coupled 2/2 chain drive	£12	£10	£7	£5
No 24 Fiat 60 HP 1905. Purple, grey, gold trim, white hood folded down; black wheels; open tourer	£15	£14	£12	£10
No 25 Fiat 24 HP Double Phaeton. Orange body, white top; black interior, gold trim, front motor 4 cyl vertical bibloc cmc 7363 hp 47 at 1200 tmp, chain transmission	£15	£13	£10	£7
No 26 Fiat 12 HP 1902. Brown, yellow wheels, blue interior, gold trim, 4 cyl motor bibloc cmc 3770 hp, 14 to 1200 tmp consumption 1 liter for every mph 4.5	£10	£9	£7	£5
No 27 Fiat 24 HP Limousine 1905. Green, red, gold trim, red wheels, 4 cyl motor cmc 7263 HP, 47 at 1200 tpm, chain transmission, red roof, black interior	£12	£10	£7	£5
No 28 Bianchi 20-30 HP 1905 Landaulet 4 Seater. Blue, black top, white hood, gold trim, orange wheels, 4 cyl motor in line cmc 4939, normal speed tpm, 1250 hp, 25 about cardanic transmission	£10	£9	£7	£5
No 29 Mercedes Simplex 1929. Yellow; brown hood, black interior, black wheels, 4 cyl, motor 28-32 hp, 40 mph	£10	£9	£7	£5
No 30 De Dion Bouton "Victoria" 1894. Steam car, trailer, car gold striped, black trim, red wheels, trailer green, red, black, gold trim, one of the first ever	£10	£9	£7	£5
No 31 Fiat 8 HP 1901. Golden yellow, black, purple trim, 2 cyl, motor boring strokex 100, total, capt 1100	£10	£8	£6	£4
No 32 Fiat 16-24 HP 1903. Red, cream roof, gold trim, red wheels, 4 cyl motor biblock-boring stroke, 110x110 total capacity 4181, a neat little car	£9	£8	£5	£3
No 33 Mercedes 1908 Limousine 70 HP. Red, black, gold trim, orange wheels, green and yellow cases, 6 cyl	£15	£12	£10	£7
No 34 Renault Type X 1907 Double Sedan. Green, gold, orange trim, red wheels, made for special travel.	£15	£14	£12	£10
No 35 Renault 1933 Model. Red, black, yellow, gilt trim; Fiacre (Marne Taxi); a gem and rare in series	£15	£14	£12	£10
No 36 Bugatti Royal Mod 41. Cream, black roof, white, red wheels, green interior, 1927, 8 cyl	£18	£15	£12	£10
No 37 Buggatti Royale Mod 41. Green; red, cream wheels, gold seats, open tourer, 8 cyl, 1927	£18	£15	£12	£10
No 38 Fiat 18/24 HP 1908. Gold, black, red strip; golden trim, red wheels, roof rack, blue running B	£12	£10	£7	£5
No 39 Rolls Royce Phantom II 1931. Blue, green, white hood, silver trim, white rim wheels, dickie seat, 6 cyl	£15	£14	£12	£10
No 40 Rolls Royce Phantom II 1931 Open. Yellow, green, silver trim, white wall tyres, brown interior, silver spokes, 6 cy	£15	£12	£10	£8
No 41 Lancia "Dilambda" V8. Rich green, black roof, black line trim, gold, silver trim, 8 cyl, 1929	£18	£15	£12	£10
No 42 Lancia "Dilambda" Torpedo 1929. Red, white or cream cover on roof, gold trim, V8 cyl	£15	£14	£12	£10
No 43 Lincoln Continental 1941. Blue, white hood; silver trim, white wall tyres, white interior, V12 cyl	£18	£15	£12	£10
No 44 Lincoln Continental V12 Open 1941. Silver grey, silver trim; white wall tyres, red interior, 12 cyl	£18	£15	£12	£10
No 45 Duesenberg "SJ" Torpedo Phaeton 1934. Blue, white hood, silver trim, spoked wheels, white rims	£18	£15	£12	£10
No 46 Duesenberg "SJ" Phaeton 1934 Open. Silver grey, silver trim, spoked white rimmed wheels, large head lights, one of the treasure of the Italian gentry	£18	£15	£12	£10
No 47 Thomas Flyabout 1908. White, silver trim, dark red wheels; black tyres, green interior; USA flag decal, rally car New York to Paris	£12	£10	£7	£5
No 48 Buggatti 5000 cc Model T50 1932. Rich red, black; white trim; silver parts, blue interior, opening doors and lift off bonnet, beautiful car	£18	£15	£12	£10
No 49 Super Fiat 12 V Dorsey De Ville 1921. Rich blue, black on top part of body, silver trim; brown wheels, open cab, silver windscreen open, one of the rare models to collect as soon as possible	£18	£16	£13	£10